Akram Louiz

A study of the Theta Functions Nullwert Forms with New Transformations

Akram Louiz

A study of the Theta Functions Nullwert Forms with New Transformations

A study of Theta Functions in their Nullwert Forms with a Proof of New Transformations of Theta-3 and Theta-4 functions

LAP LAMBERT Academic Publishing

Imprint

Cover image: www.ingimage.com

Publisher:
LAP LAMBERT Academic Publishing
is a trademark of
Dodo Books Indian Ocean Ltd. and OmniScriptum S.R.L publishing group

120 High Road, East Finchley, London, N2 9ED, United Kingdom
Str. Armeneasca 28/1, office 1, Chisinau MD-2012, Republic of Moldova, Europe
Managing Directors: Ieva Konstantinova, Victoria Ursu
info@omniscriptum.com

Printed at: see last page
ISBN: 978-620-9-05153-1

Author:

A.LOUIZ

The Higher Institute of Maritime Studies, Morocco – Naval engineering, Casablanca

[redacted]

TABLE OF CONTENTS:

1. *A discussion of a rationality and continuity related to Theta Functions by using the summation of Poisson*

Abstract:

Mathematical philosophy is built by defining logical concepts such as the concept of infinity and the concept of continuity. Hence, mathematicians are not those who can calculate, but those who can understand the mathematical concepts and can develop them in order to make new mathematical findings or to transform an hypothesis into a philosophical truth. However, mathematicians are also alarmed scientists who can verify doubted formulas by disproving a mathematical proof or finding a counterexample. It is therefore false to think that mathematicians can't surprise nowadays with new findings.

This is a mathematical demonstration and discussion concerning a special case of functions related to Theta Functions in their nullwert form and it is a logical opportunity for those who study the summation of Poisson. This work starts by using a formula of Poisson in order to make a mathematical representation for the studied function then a new approximation of this case of functions is demonstrated. Furthermore, we discuss the consequences if this studied function gives only irrationals as outputs then we use the definition of the continuity of real functions in order to make a demonstration for the continuity of the same Function. Finally, the lovers of mathematics are invited to understand the demonstrations of this article and the discussions of the continuity contradictions in order to develop new findings related to Theta Functions in their nullwert form and the summation of Poisson.

Keywords. Series, summation, Poisson, Theta function, nullwert form, rational, irrational, limit, logic, approximation, calculus, analysis.

---

1. Introduction:

It is a fact that all scientists and engineers need mathematicians. However, many principles of discrete mathematics are not well accepted among people who only apply mathematical tools. I made a previous short note of calculus about the distances on the line of real numbers by considering the example of a runner trying to run in a field which can be integrated by respecting Riemann's definition of the integrals [1]. Here is a new mathematical work that uses the summation of Poisson which is commonly applied in many fields especially for the development of the Zeta Function of Riemann. This work tests if there is a contradiction of continuity that concerns also Riemann Zeta Function and that can be annexed to the previous article that proposes an easy criterion for the study of the zeros of Riemann's hypothesis [2]. This work is dedicated therefore to pure mathematics after my previous works dedicated to mathematical physics like my thesis that proposes new mathematical tools by explaining that physics fields need to apply mathematics differently [3].

This work discusses the rationality of a sum related to Theta-functions in their nullwert form without using any long and complicated methods similar to the study of the periodicity of the binary expansion. A mathematical representation and a generalization is made for $f(B)=\sum_{n\in IN}\left(\frac{1}{B}\right)^{n^2}$ and $B\geq 2$ which defines a special function related to Theta Functions in their nullwert form. In this article, we propose a new approximation for the output values of this function after attempting to find the nature of these outputs. Then we prove the continuity of the same function $f(x)=\sum_{N\in IN}\left(\frac{1}{x}\right)^{N^2}$ when x is a real number with $x>2$.

Since it is very difficult to find a real number B where $\sum_{n\in IN}\left(\frac{1}{B}\right)^{n^2}$ is a rational number, then this short work is very useful for many fields of applied mathematics and especially for the study of Theta-functions [4], the summation of Poisson and for the study of the related Riemann Zeta function. This work can also be completed by the article that proposes new expressions for the studied series [5] which makes the findings of this paper more interesting.

2. Representation of the series to understand the rationality:

The series $\sum_{n\in IN}\left(\frac{1}{2}\right)^{n}$ is a sum of terms of unity divided by a power of 2 and everybody knows that $\sum_{n\in IN}\left(\frac{1}{2}\right)^{n}=2$ which is a rational number. Therefore, an idea that may seem rational for many students is to think that $\sum_{n\in IN}\left(\frac{1}{2}\right)^{n^2}$ is maybe a rational number or, at least, this sum is only a recurrent decimal since it is also a sum of terms of unity divided by a power of 2. However, the representation of these kind of series by using the sum of Poisson makes it difficult to decide if such series have a rational limit or not.

Let's start by using the following formula of Poisson which is also used for the proof of Riemann Zeta Function in order to make a different representation for the series $\sum_{n\in IN}\left(\frac{1}{2}\right)^{n^2}$.

The formula of Poisson is: $\sum_{n\in Z}\exp\left(-\pi\times n^2\times s\right)=\frac{1}{\sqrt{s}}\times\sum_{n\in Z}\exp\left(\frac{-\pi\times n^2}{s}\right)$ (for any : s >0) (1)

Let's consider that ***t>1*** and let's use ***s=ln(t)*** in this formula. (2)

We have :

$$\sum_{n\in Z}\exp\left(-\pi\times n^2\times\ln(t)\right)=\frac{1}{\sqrt{\ln(t)}}\times\sum_{n\in Z}\exp\left(\frac{-\pi\times n^2}{\ln(t)}\right) \quad (3)$$

$$\Leftrightarrow\sum_{n\in Z}t^{-\pi\times n^2}=\frac{1}{\sqrt{\ln(t)}}\times\sum_{n\in Z}\exp\left(\frac{-\pi\times n^2}{\ln(t)}\right) \quad (4)$$

Now let's consider that : $u=\frac{1}{t}$ and we have consequently : $0<u<1$. (5)

And we have :

$$2\times\sum_{n\in IN}u^{\pi\times n^2}-1=\frac{1}{\sqrt{-\ln(u)}}\times\sum_{n\in Z}\exp\left(\frac{\pi\times n^2}{\ln(u)}\right) \quad (6)$$

$$\text{Hence : }\sum_{n\in IN}u^{\pi\times n^2}=\frac{1}{2\times\sqrt{-\ln(u)}}\times\left(\sum_{n\in Z}\exp\left(\frac{\pi\times n^2}{\ln(u)}\right)\right)+\frac{1}{2} \quad (7)$$

Now let's consider that : $u = \left(\frac{1}{2}\right)^{\frac{1}{\pi}}$ (8)

We can verify that : 0<u<1 .

Consequently, we get this formula :

$$\sum_{n \in IN} \left(\frac{1}{2}\right)^{n^2} = \frac{1}{2 \times \sqrt{\frac{\ln(2)}{\pi}}} \times \left(\sum_{n \in Z} \exp\left(\frac{-\pi^2 \times n^2}{\ln(2)} \right) \right) + \frac{1}{2} \quad (9)$$

And thus, it is easy to remark that only a well developed mathematical demonstration can prove if $\sum_{n \in IN} \left(\frac{1}{2}\right)^{n^2}$ is rational or not.

Also, we can generalize the result above by considering a real number $B \geq 2$ when applying the formula of Poisson with : $u = \left(\frac{1}{B}\right)^{\frac{1}{\pi}}$. And we get consequently:

$$\sum_{n \in IN} \left(\frac{1}{B}\right)^{n^2} = \frac{1}{2 \times \sqrt{\frac{\ln(B)}{\pi}}} \times \left(\sum_{n \in Z} \exp\left(\frac{-\pi^2 \times n^2}{\ln(B)} \right) \right) + \frac{1}{2} \quad (10)$$

In order to have $\sum_{n \in IN} \left(\frac{1}{B}\right)^{n^2}$ rational, the coefficient $\frac{1}{2 \times \sqrt{\frac{\ln(B)}{\pi}}}$ and the series $\sum_{n \in Z} \exp\left(\frac{-\pi^2 \times n^2}{\ln(B)} \right)$ should be both rational or at least both irrational. However, even that coefficient seems to be rarely rational. Hence, it is even difficult to find a real number ***B*** which respects the fact that: $\sum_{n \in IN} \left(\frac{1}{B}\right)^{n^2}$ is rational.

Furthermore, as it is demonstrated in a personal different article before [5], this function expressed as series $f(B)=\sum_{n=1}^{n=+\infty}\left(\frac{1}{B^{n^2}}\right)$ can be expressed also for $B \geq 2$ with :

$$\sum_{n=1}^{n=+\infty}\left(\frac{1}{B^{n^2}}\right)=-\sum_{i=0}^{i=+\infty}\left(\sum_{n=1}^{n=+\infty}\left(\frac{(-1)^n}{\left(B^{2^{2i}\times n^2}\times 2^{-i}\right)}\right)\right)=\frac{1-B}{B}\times\lim_{k\to+\infty}\left(\left(\frac{1}{B}\right)^k\times\sum_{i=0}^{i=k}\left(\sum_{n=1}^{n=+\infty}\left(\frac{(-1)^n}{\left(B^{2^{2i}\times n^2-i}\times 2^{-i}\right)}\right)\right)\right) \quad (11)$$

Without forgetting that the notation «*lim* » is maintained in the formula (11) since the increasing variable **k** should tend to the same value for all the terms even in the «*infinity*».

It is also useful to explore the transcendence demonstrations of the Professor Daniel Bertrand in his article 'Theta functions and transcendence' [4], where he links the Theta-3 Functions to these kind of series as follows:

$$\theta_3(q)=1+2\times\sum_{n\geq 1}q^{n^2}=\left(2\times\sum_{n\in IN}(q)^{n^2}\right)-1=\left(2\times\sum_{n\in IN}\left(\frac{1}{B}\right)^{n^2}\right)-1 \quad \text{with : } B=\frac{1}{q} \quad (12)$$

Unfortunately, the representations of the studied series by the formula (11) doesn't seem to simplify the study of rationality of the series limit. However, we will make an attempt to find new information about the rationality by using the formula of Poisson.

3. *An attempt to explore the rationality of the studied series:*

We have: $$\sum_{n\in IN}\left(\frac{1}{B}\right)^{n^2}=\frac{1}{2\times\sqrt{\frac{\ln(B)}{\pi}}}\times\left(\sum_{n\in Z}\exp\left(\frac{-\pi^2\times n^2}{\ln(B)}\right)\right)+\frac{1}{2} \quad \text{where } B\geq 2 \quad (13)$$

If $\sum_{n\in IN}\left(\frac{1}{B}\right)^{n^2}$ is rational then we have:

$$\sum_{n\in IN}\left(\frac{1}{B}\right)^{n^2}=\frac{1}{2\times\sqrt{\frac{\ln(B)}{\pi}}}\times\left(\sum_{n\in Z}\exp\left(\frac{-\pi^2\times n^2}{\ln(B)}\right)\right)+\frac{1}{2}=\frac{p}{q} \quad \text{with } p \text{ and } q \in \{0\}. \quad (14)$$

And if the coefficient: $\frac{1}{2\times\sqrt{\frac{\ln(B)}{\pi}}}$ is rational then we have the value: $\sqrt{\frac{\ln(B)}{\pi}}$ is rational.

This means that if the coefficient: $\dfrac{1}{2\times\sqrt{\dfrac{\ln(B)}{\pi}}}$ is rational then:

$$\sqrt{\frac{\ln(B)}{\pi}}=\frac{a}{b}\quad a \text{ and } b \in \quad \{0\} \tag{15}$$

consequently: $B=\exp\left(\dfrac{a^2}{b^2}\times\pi\right)$ (16)

Now let's consider that: $\displaystyle\sum_{n\in IN}\left(\frac{1}{B}\right)^{n^2}$ is rational and that: $\dfrac{1}{2\times\sqrt{\dfrac{\ln(B)}{\pi}}}$ is also rational :

We have in this case: $\displaystyle\sum_{n\in Z}\exp\left(\frac{-\pi^2\times n^2}{\ln(B)}\right)=2\times\sum_{n\in IN}\exp\left(\frac{-\pi^2\times n^2}{\ln(B)}\right)-1$ (17)

And we have: $\displaystyle\sum_{n\in IN}\exp\left(\frac{-\pi^2\times n^2}{\ln(B)}\right)=\frac{1}{2}\times\sum_{n\in Z}\exp\left(\frac{-\pi^2\times n^2}{\ln(B)}\right)+\frac{1}{2}$ (18)

And also: $\displaystyle\sum_{n\in Z}\exp\left(\frac{-\pi^2\times n^2}{\ln(B)}\right)=2\sqrt{\frac{\ln(B)}{\pi}}\times\left(\frac{p}{q}-\frac{1}{2}\right)=\frac{2\times a}{b}\times\left(\frac{p}{q}-\frac{1}{2}\right)$ (19)

And thus, we have: $\displaystyle\sum_{n\in IN}\exp\left(\frac{-\pi^2\times n^2}{\ln(B)}\right)=\frac{a}{b}\times\left(\frac{p}{q}-\frac{1}{2}\right)+\frac{1}{2}$ (20)

However, we can notice that:

$$\sum_{n\in IN}\exp\left(\frac{-\pi^2\times n^2}{\ln(B)}\right)=\sum_{n\in IN}\frac{1}{\exp\left(\dfrac{\pi^2\times n^2}{\ln(B)}\right)}=\sum_{n\in IN}\left(\frac{1}{\exp\left(\dfrac{\pi^2}{\ln(B)}\right)}\right)^{n^2} \tag{21}$$

Hence, if we consider that : $B'=\exp\left(\dfrac{\pi^2}{\ln(B)}\right)$ (22)

Then we have $B'=\exp\left(\dfrac{\pi^2}{\ln(B)}\right)>1$ since $\ln(B)>0$. (23)

And thus we can consider that: $\sum_{n\in IN}\left(\frac{1}{\exp\left(\frac{\pi^2}{\ln(B)}\right)}\right)^{n^2}=\sum_{n\in IN}\left(\frac{1}{B'}\right)^{n^2}$ (24)

And we can use the formula of Poisson again. Hence we get:

$$\sum_{n\in IN}\exp\left(\frac{-\pi^2\times n^2}{\ln(B)}\right)=\sum_{n\in IN}\left(\frac{1}{B'}\right)^{n^2}=\frac{1}{2\times\sqrt{\frac{\ln\left(\exp\left(\frac{\pi^2}{\ln(B)}\right)\right)}{\pi}}}\times\left(\sum_{n\in Z}\exp\left(\frac{-\pi^2\times n^2}{\ln\left(\exp\left(\frac{\pi^2}{\ln(B)}\right)\right)}\right)\right)+\frac{1}{2} \quad (25)$$

And thus, we get: $\sum_{n\in IN}\exp\left(\frac{-\pi^2\times n^2}{\ln(B)}\right)=\frac{1}{2\times\sqrt{\frac{\pi}{\ln(B)}}}\times\left(\sum_{n\in Z}\exp\left(-n^2\times\ln(B)\right)\right)+\frac{1}{2}$ (26)

And since we considered that: $B=\exp\left(\frac{a^2}{b^2}\times\pi\right)$ then we get:

$$\sum_{n\in IN}\exp\left(\frac{-\pi^2\times n^2}{\ln(B)}\right)=\frac{1}{2\times\frac{b}{a}}\times\left(\sum_{n\in Z}\exp\left(\frac{-n^2\times a^2}{b^2}\times\pi\right)\right)+\frac{1}{2} \quad (27)$$

We calculated above that: $\sum_{n\in IN}\exp\left(\frac{-\pi^2\times n^2}{\ln(B)}\right)=\frac{a}{b}\times\left(\frac{p}{q}-\frac{1}{2}\right)+\frac{1}{2}$ (28)

Hence: $\frac{1}{2\times\frac{b}{a}}\times\left(\sum_{n\in Z}\exp\left(\frac{-n^2\times a^2}{b^2}\times\pi\right)\right)+\frac{1}{2}=\frac{a}{b}\times\left(\frac{p}{q}-\frac{1}{2}\right)+\frac{1}{2}$ (29)

Consequently: $\sum_{n\in Z}\exp\left(\frac{-n^2\times a^2}{b^2}\times\pi\right)=\frac{2p}{q}-1$ (30)

And we know that: $\sum_{n\in Z}\exp\left(\frac{-n^2\times a^2}{b^2}\times\pi\right)=2\times\sum_{n\in IN}\exp\left(\frac{-n^2\times a^2}{b^2}\times\pi\right)-1$ (31)

And thus we have in this case: $\sum_{n\in IN}\exp\left(\frac{-n^2\times a^2}{b^2}\times\pi\right)=\frac{p}{q}$ (32)

Consequently, if we consider that: $\sum_{n\in IN}\left(\frac{1}{B}\right)^{n^2}$ is rational and that: $\frac{1}{2\times\sqrt{\frac{\ln(B)}{\pi}}}$ is rational, then we

prove that: $\sum_{n\in IN}\left(\frac{1}{B}\right)^{n^2}=\frac{1}{2\times\sqrt{\frac{\ln(B)}{\pi}}}\times\left(\sum_{n\in Z}\exp\left(\frac{-\pi^2\times n^2}{\ln(B)}\right)\right)+\frac{1}{2}=\sum_{n\in IN}\exp\left(\frac{-n^2\times a^2}{b^2}\times\pi\right)$ (33)

with: a and $b\in\{0\}$

However, we should stop here since we remark that we already have:

$$\sum_{n\in IN}\left(\frac{1}{B}\right)^{n^2}=\sum_{n\in IN}\left(\frac{1}{\exp(\ln(B))}\right)^{n^2}=\sum_{n\in IN}\left(\frac{1}{\exp\left(\frac{a^2}{b^2}\times\pi\right)}\right)^{n^2}=\sum_{n\in IN}\exp\left(\frac{-n^2\times a^2}{b^2}\times\pi\right) \quad (34)$$

And thus, we remark that we return back to the initial formula after applying the formula of Poisson twice. Consequently, if we consider that: $\sum_{n\in IN}\left(\frac{1}{B}\right)^{n^2}$ is rational and that:

$\exists a$ and $b\in\{0\}$ with $B=\exp\left(\frac{a^2}{b^2}\times\pi\right)$ then we can't prove any new formula and our calculations lead us to an infinite loop.

4. A method to approximate the limit of the series:

Since it is difficult to find sufficient information about the rationality of the series $\sum_{n\in IN}\left(\frac{1}{B}\right)^{n^2}$ then we will propose a method to approximate the value of the series.

We have: $B^{i^2}>B^{i\times M}\Leftrightarrow\frac{1}{B^{i^2}}<\frac{1}{B^{i\times M}}$ where ***M*** and ***i*** are natural numbers and $i>M$. (35)

Hence, we have: $\sum_{i=M}^{n}\frac{1}{B^{i^2}}<\sum_{i=M}^{n}\frac{1}{B^{i\times M}}$ (36)

And thus: $\sum_{i=M}^{+\infty}\frac{1}{B^{i^2}}<\sum_{i=M}^{+\infty}\frac{1}{B^{i\times M}}$ (37)

The order is strict despite the order limit theorem.

We can remark that: $\sum_{i=M}^{+\infty}\frac{1}{B^{i\times M}}=\sum_{i=M}^{+\infty}\left(\frac{1}{B^{M}}\right)^{i}$ (38)

which have the form of geometric series with ratio ***r*** : $r=\frac{1}{B^{M}}$ (39)

And that: $\sum_{i=M}^{+\infty}\left(\frac{1}{B^{M}}\right)^{i}=\sum_{i=0}^{+\infty}\left(\frac{1}{B^{M}}\right)^{i}-\sum_{i=0}^{M-1}\left(\frac{1}{B^{M}}\right)^{i}$ (40)

and it is sufficient that ***M>1*** and ***i*** variates.

And since: $\sum_{i=0}^{+\infty}\left(\frac{1}{B^{M}}\right)^{i}=\frac{B^{M}}{B^{M}-1}$ (41)

And: $\sum_{i=0}^{M-1}\left(\frac{1}{B^{M}}\right)^{i}=\frac{1-\frac{1}{B^{M^2}}}{1-\frac{1}{B^{M}}}=\frac{B^{M}-B^{M-M^2}}{B^{M}-1}$ (42)

Then we have: $\sum_{i=M}^{+\infty}\frac{1}{B^{i^2}}<\frac{B^{M}}{B^{M}-1}-\frac{B^{M}-B^{M-M^2}}{B^{M}-1}$ (43)

And thus, we get: $\sum_{i=M}^{+\infty}\frac{1}{B^{i^2}}<\frac{B^{M-M^2}}{B^{M}-1}$ (44)

Finally, we have: $\sum_{n\in IN}\left(\frac{1}{B}\right)^{n^2}<\frac{B^{M-M^2}}{B^{M}-1}+\sum_{i=0}^{M-1}\frac{1}{B^{i^2}}$ with ***M>1*** (45)

This result may be very useful for guessing the values of the limits of the series $\sum_{n\in IN}\left(\frac{1}{B}\right)^{n^2}$ which can even help to study the rationality of this series. However, the best method to use this result as an approximation is by using: $\sum_{i=0}^{M-1}\frac{1}{B^{i^2}} < \sum_{n\in IN}\left(\frac{1}{B}\right)^{n^2} < \frac{B^{M-M^2}}{B^M-1}+\sum_{i=0}^{M-1}\frac{1}{B^{i^2}}$ with ***M>1*** (46)

And generally, it is easy to conclude that for any finite natural number ***K***, we have:

$$\sum_{i=0}^{M-1}\frac{1}{B^{i^2+K}} < \sum_{n\in IN}\left(\frac{1}{B}\right)^{n^2+K} < \frac{B^{M-M^2}}{B^K\times\left(B^M-1\right)}+\sum_{i=0}^{M-1}\frac{1}{B^{i^2+K}} \quad \text{with } M>1 \tag{47}$$

And we remark that this approximation becomes better when we increase the value of the natural number ***M***.

5. Discussing the discontinuity of the studied Theta Function:

If we consider because of the aspect of the formula of Poisson that the limit of $\sum_{n\in IN}\left(\frac{1}{B}\right)^{n^2}$ is not a rational number when ***B*** is a real number with $B\geq 2$, then this means that the function $f(B)=\sum_{n\in IN}\left(\frac{1}{B}\right)^{n^2}$ when ***B*** is a real number with $B\geq 2$ will be a real function that gives only irrational numbers. We would say that the image by the function ***f*** of the positive real number line that starts from the origin 2 is only a not continuous set of points. This means therefore that this function ***f*** is not continuous at all in any point of its curve when $B\geq 2$ since Q is dense in IR. However, let's prove this result mathematically:

First, we start by proving that the function $f(B)=\sum_{n\in IN}\left(\frac{1}{B}\right)^{n^2}$ when ***B*** is a real number with $B\geq 2$ is strictly monotone. Let ***a*** and ***b*** be real numbers with $a>b\geq 2$***:***

We have: $a>b\Rightarrow\frac{1}{a}<\frac{1}{b}\Rightarrow\left(1+\frac{1}{a}\right)-\left(1+\frac{1}{b}\right)<0$ (48)

And we have: $a>b\Rightarrow\frac{1}{a^{m^2}}<\frac{1}{b^{m^2}}$ for any natural number ***m*** with ***m>1.*** (49)

Hence we have: $a>b\Rightarrow\sum_{m=2}^{M}\frac{1}{a^{m^2}}<\sum_{m=2}^{M}\frac{1}{b^{m^2}}$ for any natural number ***M with M>m.*** (50)

And thus: $a > b \Rightarrow \sum_{m=2}^{+\infty} \frac{1}{a^{m^2}} - \sum_{m=2}^{+\infty} \frac{1}{b^{m^2}} \leq 0$ (51)

And we have: $1 + \frac{1}{a} + \sum_{m=2}^{+\infty} \frac{1}{a^{m^2}} = \sum_{m=0}^{+\infty} \frac{1}{a^{m^2}}$ and: $1 + \frac{1}{b} + \sum_{m=2}^{+\infty} \frac{1}{b^{m^2}} = \sum_{m=0}^{+\infty} \frac{1}{b^{m^2}}$ (52)

Consequently: $a > b \Rightarrow \sum_{m=0}^{+\infty} \frac{1}{a^{m^2}} - \sum_{m=0}^{+\infty} \frac{1}{b^{m^2}} < 0 \Rightarrow \sum_{m=0}^{+\infty} \frac{1}{a^{m^2}} < \sum_{m=0}^{+\infty} \frac{1}{b^{m^2}}$ (53)

This proves therefore that the function $f(B) = \sum_{n \in IN} \left(\frac{1}{B}\right)^{n^2}$ when ***B*** is a real number with $B \geq 2$ is strictly monotone.

It is sufficient to remark that if the studied Function ***f*** is continuous in any closed interval included in $I = [2;+\infty[$ then this function is bijective in any closed interval included in $I = [2;+\infty[$ since we have also $f(B) = \sum_{n \in IN} \left(\frac{1}{B}\right)^{n^2}$ strictly monotone. However, if the set of outputs of this function is made only of irrationals, then rational numbers included in any closed interval included in $I = [2;+\infty[$ can never be images by this function as demonstrated above. And thus, this is a contradiction with the fact that the studied Function ***f*** is bijective in any closed interval included in $I = [2;+\infty[$. This means therefore that this function is not continuous.

We can also prove the discontinuity in this case of irrationality by using the definition of continuity in term of sequences. Let's consider that we have this sequence $A_n = \left(a - \frac{1}{n}\right)$ where we have

$\lim_{n \to +\infty} A_n = \left(a - \frac{1}{n}\right) = a$. (54)

Let's consider that $\lim_{n \to +\infty} f(A_n) = \frac{10}{7}$ and that ***a*** is included in a closed interval included in $I = [2;+\infty[$.

We chose $\frac{10}{7}$ as an example of rational number as limit of the studied function ***f*** and ***n*** is any natural strictly positive number. If you say that the studied Function ***f*** is continuous in any closed interval included in $I = [2;+\infty[$ then you should know that this function ***f*** is also strictly monotone. The studied Function ***f*** is consequently bijective in any closed interval included in $I = [2;+\infty[$ and it should respect the definition of continuity in term of sequences.

Hence, $\lim_{n \to +\infty} f(A_n) = \frac{10}{7}$ should be a consequence of $\lim_{n \to +\infty} A_n = \left(a - \frac{1}{n}\right) = f^{-1}\left(\frac{10}{7}\right)$ (55)

since this function ***f*** should take all possible values, irrational or not, in any bounded interval in its domain of definition. However, we considered that the set of output values of this studied Function is made only of irrationals. This means that there exists no point ***a*** of the domain of definition of this function with $f(a)=\frac{10}{7}$ since $f(a)$ can only be an irrational number from the set of outputs of the studied function of irrationals. And thus, we proved that $\lim_{n\to+\infty} f(A_n)=\frac{10}{7}$ can't be equal to $f(a)$ even if we considered that $\lim_{n\to+\infty} A_n = a$.

This means that: $\lim_{n\to+\infty} A_n = a$ and $\lim_{n\to+\infty} f(A_n)\neq f(a)$ and thus the definition of continuity in term of sequences can't be verified and respected for this studied Function ***f*** if the set of its ouputs is made only of irrationals.

And this means also in this case that the related Theta Function $\theta_3(q)$ is not continuous since it is defined by:

$$\theta_3(q)=1+2\times\sum_{n\geq1}q^{n^2}=\left(2\times\sum_{n\in IN}(q)^{n^2}\right)-1=\left(2\times\sum_{n\in IN}\left(\frac{1}{B}\right)^{n^2}\right)-1 \text{ with : } B=\frac{1}{q} \tag{56}$$

6. The proof of continuity of the studied Theta Function:

We can obviously remark that the curve of the function f, or of a part of it, seems continuous by observing the curve of this function on MATLAB or any other Analysis and calculus software. Let's make therefore an investigation of continuity without using the formula of Poisson.

Let's start by proving the continuity of the function $f(x)=\sum_{N\in IN}\left(\frac{1}{x}\right)^{N^2}$ at the right of the real point ***a*** with $a>2$. Let's use the mathematical definition of the continuity of real functions.

The definition of the continuity in this case is:

$$\forall\varepsilon>0\ \exists\eta>0\ \forall x\in I\quad \left(|x-a|\leq\eta\Rightarrow|f(x)-f(a)|\leq\varepsilon\right) \tag{57}$$

Hence, since x>a then the definition becomes:

$$\forall\varepsilon>0\ \exists\eta>0\ \forall x\in I\quad \left(x-a\leq\eta\Rightarrow f(a)-f(x)\leq\varepsilon\right) \tag{58}$$

And thus, we get the following definition:

$$\forall \varepsilon > 0 \;\; \exists \eta > 0 \;\; \forall x \in I \quad \left(x - a \leq \eta \Rightarrow \sum_{N \in IN} \left(\frac{1}{a}\right)^{N^2} - \sum_{N \in IN} \left(\frac{1}{x}\right)^{N^2} \leq \varepsilon \right) \tag{59}$$

Let's consider that: $$\eta' = \frac{1}{2^n} \times 2 \times (\varepsilon - 1) \times \frac{(x \times a)^{n^2}}{\sum_{i=0}^{n^2-1} x^{n^2-1-i} \times a^i} \tag{60}$$

where n can be any natural number with $n \geq 2$

Let's consider that: $$\eta = \frac{1}{2^n} \times 2 \times (\varepsilon - 1) \times \frac{(2 \times a)^{n^2}}{\sum_{i=0}^{n^2-1} 2^{n^2-1-i} \times a^i} \tag{61}$$

where n can be any natural number with $n \geq 2$.

we have: $x > 2 \Rightarrow \eta < \eta'$ (62)

Now let's verify:

We have: $$x - a \leq \eta \Rightarrow x - a \leq \eta' \Rightarrow x - a \leq \frac{1}{2^n} \times 2 \times (\varepsilon - 1) \times \frac{(x \times a)^{n^2}}{\sum_{i=0}^{n^2-1} x^{n^2-1-i} \times a^i} \tag{63}$$

Hence: $$\frac{(x-a) \times \sum_{i=0}^{n^2-1} \left(x^{n^2-1-i} \times a^i \right)}{(x \times a)^{n^2}} \leq \frac{1}{2^n} \times 2 \times (\varepsilon - 1) \tag{64}$$

And thus, we have: $$\frac{x^{n^2} - a^{n^2}}{(x \times a)^{n^2}} \leq \frac{1}{2^n} \times 2 \times (\varepsilon - 1) \tag{65}$$

Consequently, we get: $$\frac{1}{a^{n^2}} - \frac{1}{x^{n^2}} \leq \frac{1}{2^n} \times 2 \times (\varepsilon - 1) \tag{66}$$

However, we should remark also that we have: $$\sum_{n=2}^{m} \left(\frac{1}{a^{n^2}} - \frac{1}{x^{n^2}} \right) \leq \sum_{n=2}^{m} \left(\frac{1}{2^n} \times 2 \times (\varepsilon - 1) \right) \tag{67}$$

Hence: $\sum_{n=2}^{m}\frac{1}{a^{n^2}}-\sum_{n=2}^{m}\frac{1}{x^{n^2}}\leq 2\times(\varepsilon-1)\times\sum_{n=2}^{m}\frac{1}{2^n}$ (68)

Let's remark that: $\sum_{n=0}^{m}\frac{1}{a^{n^2}}-\sum_{n=0}^{m}\frac{1}{x^{n^2}}=\frac{1}{a}-\frac{1}{x}+\sum_{n=2}^{m}\frac{1}{a^{n^2}}-\sum_{n=2}^{m}\frac{1}{x^{n^2}}$ (69)

And we know that: $\frac{1}{a}-\frac{1}{x}=\frac{x-a}{a\times x}\leq 1$ (70)

Consequently: $\sum_{n=0}^{m}\frac{1}{a^{n^2}}-\sum_{n=0}^{m}\frac{1}{x^{n^2}}\leq 1+2\times(\varepsilon-1)\times\sum_{n=2}^{m}\frac{1}{2^n}$ (71)

And thus: $\sum_{n=0}^{+\infty}\frac{1}{a^{n^2}}-\sum_{n=0}^{+\infty}\frac{1}{x^{n^2}}\leq 1+2\times(\varepsilon-1)\times\sum_{n=2}^{+\infty}\frac{1}{2^n}$ (72)

And since: $\sum_{n=2}^{+\infty}\frac{1}{2^n}=\frac{1}{2}$ (73)

We conclude that: $\sum_{n=0}^{+\infty}\frac{1}{a^{n^2}}-\sum_{n=0}^{+\infty}\frac{1}{x^{n^2}}\leq \varepsilon$ (74)

This means that: $x-a\leq\eta\Rightarrow\sum_{n=0}^{+\infty}\frac{1}{a^{n^2}}-\sum_{n=0}^{+\infty}\frac{1}{x^{n^2}}\leq \varepsilon$ (75)

Finally, we conclude that we have indeed:

$$\forall\varepsilon>0\ \ \exists\eta>0\ \ \forall x\in I\ \ \left(x-a\leq\eta\Rightarrow\sum_{N\in IN}\left(\frac{1}{a}\right)^{N^2}-\sum_{N\in IN}\left(\frac{1}{x}\right)^{N^2}\leq\varepsilon\right) \tag{76}$$

With: $\eta=\frac{1}{2^n}\times 2\times(\varepsilon-1)\times\frac{(2\times a)^{n^2}}{\sum_{i=0}^{n^2-1}2^{n^2-1-i}\times a^i}$ where n can be any natural number with $n\geq 2$. (77)

This means that the function $f(x)=\sum_{N\in IN}\left(\frac{1}{x}\right)^{N^2}$ when x is a real number with $x>2$ is continuous at the right in the domain $I=]2;+\infty[$.

We can follow the same steps in order to prove the continuity of the function of irrationals f at the left of the point ***a*** which is in the domain $I=]2;+\infty[$.

Since x<a then the definition becomes:

$$\forall\varepsilon>0\ \ \exists\eta>0\ \ \forall x\in I\ \ \left(a-x\leq\eta\Rightarrow f(x)-f(a)\leq\varepsilon\right) \tag{78}$$

And thus, we get the following definition:

$$\forall\varepsilon>0\ \ \exists\eta>0\ \ \forall x\in I\ \ \left(a-x\leq\eta\Rightarrow\sum_{N\in IN}\left(\frac{1}{x}\right)^{N^2}-\sum_{N\in IN}\left(\frac{1}{a}\right)^{N^2}\leq\varepsilon\right) \tag{79}$$

We consider here that: $$\eta'=\frac{1}{2^n}\times 2\times(\varepsilon-1)\times\frac{(x\times a)^{n^2}}{\sum_{i=0}^{n^2-1}a^{n^2-1-i}\times x^i} \tag{80}$$

where n can be any natural number with $n\geq 2$

We consider also that: $$\eta=\frac{1}{2^n}\times 2\times(\varepsilon-1)\times\frac{(2\times a)^{n^2}}{\sum_{i=0}^{n^2-1}a^{n^2-1-i}\times 2^i} \tag{81}$$

where n can be any natural number with $n\geq 2$

we have: $$x>2\Rightarrow\eta<\eta' \tag{82}$$

Now let's verify:

We have: $$a-x\leq\eta\Rightarrow a-x\leq\eta'\Rightarrow a-x\leq\frac{1}{2^n}\times 2\times(\varepsilon-1)\times\frac{(x\times a)^{n^2}}{\sum_{i=0}^{n^2-1}a^{n^2-1-i}\times x^i} \tag{83}$$

Hence: $$\frac{(a-x)\times\sum_{i=0}^{n^2-1}\left(a^{n^2-1-i}\times x^i\right)}{(x\times a)^{n^2}}\le\frac{1}{2^n}\times 2\times(\varepsilon-1) \quad (84)$$

And thus, we have: $$\frac{a^{n^2}-x^{n^2}}{(x\times a)^{n^2}}\le\frac{1}{2^n}\times 2\times(\varepsilon-1) \quad (85)$$

Consequently, we get: $$\frac{1}{x^{n^2}}-\frac{1}{a^{n^2}}\le\frac{1}{2^n}\times 2\times(\varepsilon-1) \quad (86)$$

And by following the same steps, we get: $$\sum_{n=0}^{+\infty}\frac{1}{x^{n^2}}-\sum_{n=0}^{+\infty}\frac{1}{a^{n^2}}\le 1+2\times(\varepsilon-1)\times\sum_{n=2}^{+\infty}\frac{1}{2^n} \quad (87)$$

We conclude that: $$\sum_{n=0}^{+\infty}\frac{1}{x^{n^2}}-\sum_{n=0}^{+\infty}\frac{1}{a^{n^2}}\le\varepsilon \quad \text{since:} \quad \sum_{n=2}^{+\infty}\frac{1}{2^n}=\frac{1}{2} \quad (88)$$

This means that: $$a-x\le\eta\Rightarrow\sum_{n=0}^{+\infty}\frac{1}{x^{n^2}}-\sum_{n=0}^{+\infty}\frac{1}{a^{n^2}}\le\varepsilon \quad (89)$$

Finally, we conclude that we have indeed:

$$\forall\varepsilon>0\ \ \exists\eta>0\ \ \forall x\in I\ \ \left(a-x\le\eta\Rightarrow\sum_{N\in IN}\left(\frac{1}{x}\right)^{N^2}-\sum_{N\in IN}\left(\frac{1}{a}\right)^{N^2}\le\varepsilon\right) \quad (90)$$

With: $\eta=\frac{1}{2^n}\times 2\times(\varepsilon-1)\times\frac{(2\times a)^{n^4}}{\sum_{i=0}^{n^4-1}a^{n^4-1-i}\times 2^i}$ where n can be any natural number with $n\ge 2$. (91)

This means that the function $f(x)=\sum_{N\in IN}\left(\frac{1}{x}\right)^{N^2}$ when x is a real number with $x>2$ is continuous at the left in the domain $I=]2;+\infty[$.

We conclude that the function $f(x)=\sum_{N\in IN}\left(\frac{1}{x}\right)^{N^2}$ when x is a real number with $x>2$ is continuous in all the domain $I=]2;+\infty[$.

Finally, we conclude that the related Theta Function $\theta_3(q)$ is continuous in all the domain $I=]2;+\infty[$.

7. Summary and conclusion:

The article started by describing how difficult is to guess or even prove if the values of the studied function $f(B)=\sum_{n\in IN}\left(\frac{1}{B}\right)^{n^2}$ is rational or irrational. Hence, each real value ***B*** can be studied alone by using mathematical tools similar to the approximation demonstrated in this work which is:

$$\sum_{i=0}^{M-1}\frac{1}{B^{i^2+K}}<\sum_{n\in IN}\left(\frac{1}{B}\right)^{n^2+K}<\frac{B^{M-M^2}}{B^K\times\left(B^M-1\right)}+\sum_{i=0}^{M-1}\frac{1}{B^{i^2+K}}$$

where ***K*** and ***M*** are finite natural numbers with ***M>1*** and where the approximation becomes better when we increase the value of ***M***.

This work also proves the continuity of this function ***f*** with the definition after demonstrating that if the set of outputs of this function is purely irrational, or purely rational, then this contradicts the continuity of this studied function ***f***.

Finally, since it is very difficult to find a real number ***y*** where $\sum_{n\in IN}\left(\frac{1}{y}\right)^{n^2}$ is a rational number, then I still invite the readers to discuss the rationality of the function $f(B)=\sum_{n\in IN}\left(\frac{1}{B}\right)^{n^2}$ because proving mathematically its irrationality in any interval will threaten the use of the summation of Poisson which is widely applied in many fields especially for any proof related to Riemann Zeta Function.

Declarations:

Ethical Approval

this declaration is "not applicable".

Competing interests

this declaration is "not applicable".

Authors' contributions

this declaration is "not applicable".

Funding

this declaration is "not applicable".

Availability of data and materials

this declaration is "not applicable".

References:

[1] Louiz, A. (2023). Mathematics Letter: A runner cheating thanks to Riemann. *Maghrebian Journal of Pure and Applied Science*, *9*(1).

[2] akram louiz. Complex mathematical statements useful as criterion for Landau-Siegel Zeros, 20 February 2023, PREPRINT (Version 3) available at Research Square [https://doi.org/10.21203/rs.3.rs-2520944/v3]

[3] Louiz, A. (2020). A thesis about Newtonian mechanics rotations and about differential operators. Maghrebian Journal of Pure and Applied Science, 6(1), 26-50.

[4] Bertrand, D. (1997). Theta functions and transcendence. The Ramanujan Journal, 1(4), 339-350.

[5] Akram Louiz. A new equation of four infinite series and sums by using a generalized recurrence relation, 05 January 2023, PREPRINT (Version 4) available at Research Square [https://doi.org/10.21203/rs.3.rs-2378268/v4]

2. Using a theorem of Dirichlet to prove that the rational numbers as input values in Theta-3 Functions give only irrational numbers as output

Abstract:

Theta Functions are not an easy subject of mathematics because new researches about this wide topic are still being published regularly. The findings about Theta Functions lead directly to many applications in different fields of applied mathematics and help also to develop old interesting mathematical topics such as Poisson summation and Riemann Zeta Function. This work proves by using a theorem of Dirichlet and a previously demonstrated useful approximation that the rational inputs of Theta-3 Functions in their nullwert form give only irrationals as outputs of these functions. I invite consequently even the students of mathematics to understand the demonstrations of this article which are not complicated since only simple notions of sequences are being used.

Kevwords. Theta Function, nullwert form, Dirichlet, approximation, rational, irrational, theorem, continuity, sequence,

1. Introduction:

After some works of mathematical physics which try to prevent the bad application of pure mathematics in physics [1,2] or try to propose new mathematical tools for the modeling of problems in other fields[3], new works of pure mathematics have been published in order to collaborate in the solving of famous problems [4] or in order to demonstrate new formulas which can be very useful in establishing new demonstrations [5] .

However, in this article, we will carry on the efforts of the previous article [6] which tries to investigate the continuity and the rationality of a studied kind of Theta Functions in their nullwert form.

We will prove in this article that Theta-3 Functions give only irrationals as outputs when the input values of these functions are rational. We won't use in this article any complicated notions of Algebra but we will use in this proof a very useful approximation demonstrated in the previous article [6] and we will use also a famous ***Dirichlet's*** theorem about rationality.

We will use this Dirichlet's theorem which says that:

"If there are infinitely many distinct rationals p and q such that, for some strictly positive real number $\boldsymbol{\delta}$ we have: $\left|x-\frac{p}{q}\right|<\frac{1}{q^{1+\delta}}$, then $\boldsymbol{x}$ is irrational." (1)

And thus, to prove that a real number $\boldsymbol{x}$ is irrational, it is sufficient to prove that a strictly positive real number $\boldsymbol{\delta}$ exists and that a rational sequence $\frac{(p_M)_{M\in IN}}{(q_M)_{M\in IN}}$ exists (2)

with $\frac{p_M}{q_M}\neq x$ and $\left|x-\frac{p_M}{q_M}\right|<\frac{1}{q_M^{1+\delta}}$ for all natural numbers $\boldsymbol{M}$. (3)

Let's use that Dirichlet's theorem in order to investigate the case of the real number $x=\sum_{n\in IN}\left(\frac{1}{B}\right)^{n^2}$ where $\boldsymbol{B}$ is a rational number with $B>2$.

Let's consider that $\boldsymbol{M}$ is a natural number and that we have the following two rational sequences:

The first is: $(p_M)_{M\in IN}$ which respects:

$p_M=B+M+1=B+1$ *for* $M=0$ and $p_M=B+M=B+1$ *for* $M=1$ (4)

And respects: $p_M=\sum_{i=0}^{M-1}B^{-M-i^2}$ *for* $M\geq 2$ with $\boldsymbol{M}$ finite (5)

And respects: $\lim_{M\to+\infty}p_M=\lim_{M\to+\infty}\left(\frac{1}{2\times B}+\frac{1}{M}\right)=\frac{1}{2\times B}$ (6)

The second is: $(q_M)_{M\in IN}$ which respects:

$q_M=B+M=B$ *for* $M=0$ and $q_M=B+M-1=B$ *for* $M=1$ (7)

And respects: $q_M=B^{-M}$ *for* $M\geq 2$ with $\boldsymbol{M}$ finite (8)

And respects: $\lim_{M \to +\infty} q_M = \lim_{M \to +\infty} \left(\frac{1}{2} + \frac{1}{M} \right) = \frac{1}{2}$ (9)

And thus, we have this rational sequence which respects:

$$\frac{(p_M)_{M \in IN}}{(q_M)_{M \in IN}} = \frac{B+1}{B} \quad for \quad M = 0 \text{ and } M = 1 \tag{10}$$

and: $$\frac{(p_M)_{M \in IN}}{(q_M)_{M \in IN}} = \frac{\sum_{i=0}^{M-1} B^{-M-i^2}}{B^{-M}} = \sum_{i=0}^{M-1} \frac{1}{B^{i^2}} \quad for \quad M \geq 2 \text{ and M finite} \tag{11}$$

and: $$\lim_{M \to +\infty} \frac{(p_M)_{M \in IN}}{(q_M)_{M \in IN}} = \frac{\lim_{M \to +\infty} p_M}{\lim_{M \to +\infty} q_M} = \frac{1}{B} \tag{12}$$

When $x = \sum_{n \in IN} \left(\frac{1}{B} \right)^{n^2}$ we have consequently:

$$\frac{(p_M)_{M \in IN}}{(q_M)_{M \in IN}} \neq x \text{ for all natural numbers } \boldsymbol{M}. \tag{13}$$

Let's consider also that we have a real strictly positive number δ which respects :

$$0 < \delta < \frac{\ln(B-1)}{\ln(B)} \tag{14}$$

2. The case of the finite natural numbers M with $M \geq 2$*:*

Let's start by the case of $M \geq 2$ with M finite:

We have the following approximation from a previously published article [6]:

$$\sum_{i=0}^{M-1} \frac{1}{B^{i^2}} < \sum_{n \in IN} \left(\frac{1}{B} \right)^{n^2} < \frac{B^{M-M^2}}{B^M - 1} + \sum_{i=0}^{M-1} \frac{1}{B^{i^2}} \quad \text{with } M \geq 2 \tag{15}$$

This means that: $$0 < \sum_{n\in IN}\left(\frac{1}{B}\right)^{n^2} - \sum_{i=0}^{M-1}\frac{1}{B^{i^2}} < \frac{B^{M-M^2}}{B^M - 1} \qquad (16)$$

And we have: $$\left|\sum_{n\in IN}\left(\frac{1}{B}\right)^{n^2} - \sum_{i=0}^{M-1}\frac{1}{B^{i^2}}\right| = \sum_{n\in IN}\left(\frac{1}{B}\right)^{n^2} - \sum_{i=0}^{M-1}\frac{1}{B^{i^2}} \qquad (17)$$

This means that we have: $$\left|\sum_{n\in IN}\left(\frac{1}{B}\right)^{n^2} - \frac{(p_M)_{M\in IN}}{(q_M)_{M\in IN}}\right| < \frac{B^{M-M^2}}{B^M - 1} \quad \text{with } M \geq 2 \text{ and } \boldsymbol{M} \text{ finite} \qquad (18)$$

Since $\boldsymbol{B > 2}$ and $M \geq 2$ we have: $$\frac{1}{q_M^{(1+\delta)}} = B^{M\times(1+\delta)} > 1 \qquad (19)$$

And since $\boldsymbol{B > 2}$ and $M \geq 2$ we have: $$\frac{1}{B^{(M-1)}\times 6} < 1 \qquad (20)$$

Consequently, we have: $$\frac{1}{B^{(M-1)}\times 6} < \frac{1}{q_M^{(1+\delta)}} \qquad (21)$$

And we have: $$B > 2 \text{ and } M \geq 2 \Rightarrow 1 - \frac{1}{B^M} > 1 - \frac{1}{2^M} \geq 1 - \frac{1}{2^2} \qquad (22)$$

Hence, we have: $$B > 2 \text{ and } M \geq 2 \Rightarrow 1 - \frac{1}{B^M} > \frac{3}{4} \qquad (23)$$

And we have: $$B > 2 \text{ and } M \geq 2 \Rightarrow B^{M^2-M+1} > 2^{2^2-2+1} \qquad (24)$$

Hence, we have: $$B^{M^2-M+1} > 8 \qquad (25)$$

And thus, we have: $$B^{M^2-M+1}\times\left(1 - \frac{1}{B^M}\right) > 6 \qquad (26)$$

And since: $$B^{M-1}\times B^{M^2-M+1}\times\left(1 - \frac{1}{B^M}\right) = B^{M^2} - B^{M^2-M} \qquad (27)$$

Then we have: $$B^{M^2} - B^{M^2-M} > 6\times B^{M-1} \qquad (28)$$

And this means that: $$\frac{1}{B^{M^2} - B^{M^2-M}} < \frac{1}{6\times B^{M-1}} \qquad (29)$$

However, we proved that: $$\frac{1}{6\times B^{M-1}} < \frac{1}{q_M^{(1+\delta)}} \qquad (30)$$

And thus, we proved that: $$\frac{1}{B^{M^2} - B^{M^2-M}} < \frac{1}{q_M^{(1+\delta)}} \qquad (31)$$

And this is equivalent to: $\frac{B^{M-M^2}}{B^M-1}<\frac{1}{q_M^{(1+\delta)}}$ (32)

And thus, we proved that for ***B*** is a rational number with $B>2$ and for ***M*** is a finite natural number with $M\geq 2$ and for ***δ*** is a real number with $0<\delta<\frac{\ln(B-1)}{\ln(B)}$ we have:

$$0<\sum_{n\in IN}\left(\frac{1}{B}\right)^{n^2}-\sum_{i=0}^{M-1}\frac{1}{B^{i^2}}<\frac{B^{M-M^2}}{B^M-1}<\frac{1}{q_M^{(1+\delta)}} \quad (33)$$

Consequently, we proved that for ***B*** is a rational number with $B>2$ and for ***M*** is a finite natural number with $M\geq 2$ and for ***δ*** is a real number with $0<\delta<\frac{\ln(B-1)}{\ln(B)}$ we have:

$$\left|\sum_{n\in IN}\left(\frac{1}{B}\right)^{n^2}-\frac{(p_M)_{M\in IN}}{(q_M)_{M\in IN}}\right|<\frac{1}{q_M^{(1+\delta)}} \quad (34)$$

3. The case of the natural numbers M with M=0 or M=1:

Now let's study the case of ***M=0*** and ***M=1:***

We have: $p_M=B+1$ and $q_M=B$ (35)

In this case, we have: $\left|\sum_{n\in IN}\left(\frac{1}{B}\right)^{n^2}-\frac{p_M}{q_M}\right|=\left|\sum_{n\in IN}\left(\frac{1}{B}\right)^{n^2}-\frac{B+1}{B}\right|$ (36)

And since: $\sum_{n\in IN}\left(\frac{1}{B}\right)^{n^2}>1+\frac{1}{B}=\frac{B+1}{B}$ (37)

we have: $\left|\sum_{n\in IN}\left(\frac{1}{B}\right)^{n^2}-\frac{p_M}{q_M}\right|=\sum_{n\in IN}\left(\frac{1}{B}\right)^{n^2}-\frac{B+1}{B}$ (38)

And we know that: $\sum_{n\in IN}\left(\frac{1}{B}\right)^{n^2} < \sum_{n\in IN}\left(\frac{1}{B}\right)^{n}$ (39)

where: $\sum_{n\in IN}\left(\frac{1}{B}\right)^{n} = \frac{B}{B-1}$ (40)

Consequently, we have: $\sum_{n\in IN}\left(\frac{1}{B}\right)^{n^2} - \frac{p_M}{q_M} < \frac{B}{B-1} - \frac{B+1}{B}$ (41)

And we have: $\frac{1}{q_M^{(1+\delta)}} = \frac{1}{B^{1+\delta}}$ (42)

Hence: $\left(\frac{B}{B-1} - \frac{B+1}{B}\right)\times B^{1+\delta} = \frac{B^{1+\delta}}{B\times(B-1)} = \frac{B^{\delta}}{B-1}$ (43)

We have also: $\delta < \frac{\ln(B-1)}{\ln(B)}$ (44)

Consequently: $\delta\times\ln(B) < \ln(B-1) \Rightarrow B^{\delta} < B-1 \Rightarrow \frac{B^{\delta}}{B-1} < 1$ (45)

And thus, we have: $\left(\frac{B}{B-1} - \frac{B+1}{B}\right)\times B^{1+\delta} < 1$ (46)

Hence, we have: $\frac{B}{B-1} - \frac{B+1}{B} < \frac{1}{q_M^{(1+\delta)}}$ (47)

We conclude that: $\sum_{n\in IN}\left(\frac{1}{B}\right)^{n^2} - \frac{p_M}{q_M} < \frac{1}{q_M^{(1+\delta)}}$ (48)

And this means that: $\left|\sum_{n\in N}\left(\frac{1}{B}\right)^{n^2} - \frac{p_M}{q_M}\right| < \frac{1}{q_M^{(1+\delta)}}$ (49)

Consequently, we proved that for ***B*** is a rational number with $B > 2$ and for ***M*** is a finite natural number with ***M=0*** or ***M=1*** and for ***δ*** is a real number with $0 < \delta < \frac{\ln(B-1)}{\ln(B)}$ we have:

$$\left|\sum_{n\in IN}\left(\frac{1}{B}\right)^{n^2} - \frac{(p_M)_{M\in IN}}{(q_M)_{M\in IN}}\right| < \frac{1}{q_M^{(1+\delta)}} \quad (50)$$

4. The case of natural numbers M approaching the infinity:

In order to respect the fact that $\frac{(p_M)_{M\in IN}}{(q_M)_{M\in IN}} \neq x$ for all natural numbers ***M,*** let's study also the case when ***M*** approaches infinity:

Let's consider that: $\lim\limits_{M\to+\infty} p_M = \lim\limits_{M\to+\infty}\left(\frac{1}{2\times B}+\frac{1}{M}\right)=\frac{1}{2\times B}$ and: $\lim\limits_{M\to+\infty} q_M = \lim\limits_{M\to+\infty}\left(\frac{1}{2}+\frac{1}{M}\right)=\frac{1}{2}$ (51)

Let's accept in this case to write: $\lim\limits_{M\to+\infty} p_M = p_M$ and: $\lim\limits_{M\to+\infty} q_M = q_M$.

Since ***M*** approaches infinity, we have the following limits: $p_M = \frac{1}{2\times B}$ and $q_M = \frac{1}{2}$ (52)

In this case, we have: $\left|\sum\limits_{n\in IN}\left(\frac{1}{B}\right)^{n^2} - \frac{p_M}{q_M}\right| = \left|\sum\limits_{n\in IN}\left(\frac{1}{B}\right)^{n^2} - \frac{1}{B}\right|$ (53)

And since: $\sum\limits_{n\in IN}\left(\frac{1}{B}\right)^{n^2} > \frac{1}{B}$ we have: $\left|\sum\limits_{n\in IN}\left(\frac{1}{B}\right)^{n^2} - \frac{p_M}{q_M}\right| = \sum\limits_{n\in IN}\left(\frac{1}{B}\right)^{n^2} - \frac{1}{B}$ (54)

Furthermore, we have: $\sum\limits_{n\in IN}\left(\frac{1}{B}\right)^{n^2} < \sum\limits_{n\in IN}\left(\frac{1}{B}\right)^{n} \leq \sum\limits_{n\in IN}\left(\frac{1}{2}\right)^{n}$ (55)

And we have: $\sum\limits_{n\in IN}\left(\frac{1}{2}\right)^{n} = 2$ (56)

And thus, we have: $\sum\limits_{n\in IN}\left(\frac{1}{B}\right)^{n^2} < 2$ (57)

And this means that: $\left|\sum\limits_{n\in IN}\left(\frac{1}{B}\right)^{n^2} - \frac{p_M}{q_M}\right| < 2 - \frac{1}{B} < 2$ (58)

And since: $\frac{1}{1+\delta} < 1 \Rightarrow \frac{1}{2} < \frac{1}{2^{\frac{1}{1+\delta}}}$ (59)

Then we have in this case: $q_M < \frac{1}{2^{\frac{1}{1+\delta}}}$ because: $q_M = \frac{1}{2}$ (60)

And this means that: $q_M^{1+\delta} < \frac{1}{2}$ (61)

Hence we proved that: $\frac{1}{q_M^{1+\delta}} > 2$ (62)

And since we proved that: $\left| \sum_{n \in IN} \left(\frac{1}{B}\right)^{n^2} - \frac{p_M}{q_M} \right| < 2$ (63)

Then we conclude that: $\left| \sum_{n \in IN} \left(\frac{1}{B}\right)^{n^2} - \frac{p_M}{q_M} \right| < \frac{1}{q_M^{1+\delta}}$ (64)

Consequently, we proved that for ***B*** is a rational number with $B > 2$ and for ***M*** is a natural number that approaches the infinity and for ***δ*** is a real number with $0 < \delta < \frac{\ln(B-1)}{\ln(B)}$ we have also:

$$\left| \sum_{n \in IN} \left(\frac{1}{B}\right)^{n^2} - \frac{(p_M)_{M \in IN}}{(q_M)_{M \in IN}} \right| < \frac{1}{q_M^{(1+\delta)}} \quad (65)$$

5. *Conclusion:*

We conclude that for ***B*** is a rational number with $B > 2$ and for any natural number ***M*** we have:

$\left| \sum_{n \in IN} \left(\frac{1}{B}\right)^{n^2} - \frac{(p_M)_{M \in IN}}{(q_M)_{M \in IN}} \right| < \frac{1}{q_M^{(1+\delta)}}$ where it is sufficient that ***δ*** is a real number with $0 < \delta < \frac{\ln(B-1)}{\ln(B)}$.

We have always $\frac{\ln(B-1)}{\ln(B)} > 0$ even when the number ***B*** approaches the infinity, consequently the real number ***δ*** always exists with ***δ>0.***

And thus, we respected the Dirichlet's theorem described above and we finally conclude that:

$x = \sum_{n \in IN} \left(\frac{1}{B}\right)^{n^2}$ is irrational for any rational number ***B*** with $B > 2$.

And since we can link the Theta-3 Functions to these kind of series as follows:

$$\theta_3(q) = 1 + 2 \times \sum_{n \geq 1} q^{n^2} = \left(2 \times \sum_{n \in IN} (q)^{n^2} \right) - 1 = \left(2 \times \sum_{n \in IN} \left(\frac{1}{B} \right)^{n^2} \right) - 1 \quad \text{with : } B = \frac{1}{q} \qquad (66)$$

Then we can conclude that Theta-3 Functions $\theta_3(q)$ in their nullwert form give only irrational output values for any rational number ***q*** as input with $0 < q < \frac{1}{2}$.

This result which can be linked to the field of Calculus and Analysis respects the findings of the professor D. Bertrand [7] who could prove in the field of Algebra that all the algebraic inputs of Theta-3 Functions in their nullwert form give only irrationals as outputs in this kind of functions.

We could make this proof thanks to the useful approximation which has been demonstrated in the previous article about the rationality of Theta functions [6] and which is:

$$\sum_{i=0}^{M-1} \frac{1}{B^{i^2}} < \sum_{n \in IN} \left(\frac{1}{B} \right)^{n^2} < \frac{B^{M-M^2}}{B^M - 1} + \sum_{i=0}^{M-1} \frac{1}{B^{i^2}} \quad \text{For any finite natural number } \boldsymbol{M} \text{ with } M \geq 2 \qquad (67)$$

We can add that for the function $f(B) = \sum_{n \in IN} \left(\frac{1}{B} \right)^{n^2}$ where ***B>2***, if the input values ***B*** of this function are irrationals and cause also the fact that the outputs of this function in this case are irrationals then we will fall in the same contradiction of continuity described in the previous article [6]. This means in this case that we will suspect the truth of many mathematical definitions.

Declarations:

Ethical Approval:

this declaration is "not applicable".

Competing interests:

this declaration is "not applicable".

Authors' contributions:

this declaration is "not applicable".

Funding:

this declaration is "not applicable".

Availability of data and materials:

this declaration is "not applicable".

References:

[1] Louiz, A. (2020). A thesis about Newtonian mechanics rotations and about differential operators. Maghrebian Journal of Pure and Applied Science, 6(1), 26-50.

[2] Louiz, A. (2023). Mathematics Letter: A runner cheating thanks to Riemann. *Maghrebian Journal of Pure and Applied Science*, *9*(1).

[3] Akram Louiz, "A Proof That the Set of NP-problems is Bigger Than the Set of P-problems by Using a Logical Consideration," WSEAS Transactions on Computers, vol. 22, pp. 159-170, 2023, DOI:10.37394/23205.2023.22.19

[4] Louiz, A. The study of cases of fixed points of a complex function depending on the inputs of this function.

[5] Louiz, A. (2024). A new equation of four infinite series and sums by using a generalized recurrence relation.

[6] Akram Louiz. A discussion of rationality and continuity related to Theta Functions by using the summation of Poisson. Asia Mathematika, In press, ⟨10.5281/zenodo.10609879⟩. ⟨hal-04442443⟩

[7] Bertrand, D. (1997). Theta functions and transcendence. The Ramanujan Jo

3. *Transformations of Theta-3 and Theta-4 functions by using a new equation of four infinite series and sums built from a generalized recurrence relation*

Abstract Motivated by the fundamental presence of mathematics in all engineering and scientific fields, this mathematical article is made in order to give a new important result about Theta functions. This is a mathematical article that starts by proving a recurrence relation that is very useful for the demonstrations of the present article. The article proves afterwards that four infinite series are equivalent and this result has been tested successfully on the softwares : « GEOGEBRA » and « MATLAB ». Consequently, new useful and simple transformations of Theta-3 and Theta-4 functions in their « Nullwert » forms have been proven.

And thus, this article opens new opportunities to demonstrate and develop new mathematical findings and observations especially in the field of Theta Functions and Zeta Functions and the other related subjects.

Keywords Series, sum, Theta Functions, recurrence, induction, equation, new, formula, Poisson, Riemann, rational, irrational, limit, logics, mathematics, analysis.

Introduction :

The presence of mathematics is fundamental in all engineering and scientific fields. However, mathematics are still being developed since many questions stay not answered such as the well known Riemann's hypothesis.

The previously published related articles deal with the sum $\sum_{n\in IN}\left(\frac{1}{B}\right)^{n^2}$ by applying a formula of Poisson [1] or by using a theorem of Dirichlet about rationality [2]. However, more had to be made in order to to carry on demonstrating and finding results in the field of series and Theta Functions. And thus, new transformations of Theta-3 and Theta-4 functions in their « Nullwert » forms have been proven in the present work. We remind that a ***« Nullwert »*** form of Theta functions is the restriction $\theta_m(\tau) = \theta_m(\tau,0)$ of a theta function $\theta_m(\tau,z)$ where the variable τ may be a complex number in the upper half-plane with rational characteristic m to $z = 0$.

This article, started by proving a useful recurrence relation that is fundamental for the proof in this paper and also for other problems of mathematics. This formula is developed in order to produce a new infinite series that is equal to the infinite series $\sum_{n=1}^{n=+\infty}\left(\frac{1}{2}\right)^{n^2}$. This article proves also a new equation composed of four infinite equivalent series and sums. Finally, the equation of the infinite series and the other results are generalized to positive real numbers B and to the powers of natural numbers C. The results have been obviously tested successfully on the softwares : ***« GEOGEBRA »*** and ***« MATLAB »*** like it is the case of many papers of applied mathematics [3]. Consequently, new useful and simple transformations of Theta-3 and Theta-4 functions in their « Nullwert » forms have been also concluded.

This paper can be therefore very useful to the study of Theta Functions in their complex forms, and especially the Theta-3 Functions and Theta-4 Functions where the formulations explained in the works of Professor Daniel Bertrand [2,4,5] can also be very useful. And thus, we hope that this paper can give hints of new methods regarding analysis and calculus [6,7]

1. The useful recurrence relation :

In a previous article [1], we used the element $q = \frac{1}{2}$ and we made the following remark :

The series $\sum_{n \in IN}\left(\frac{1}{2}\right)^{n^2}$ is composed of some terms of the series $\sum_{n \in IN}\left(\frac{1}{2}\right)^{n}$ and since these two series are increasing monotonic positive series then :

$$\sum_{n \in IN}\left(\frac{1}{2}\right)^{n^2} \text{ is convergent and : } \sum_{n \in IN}\left(\frac{1}{2}\right)^{n^2} < \sum_{n \in IN}\left(\frac{1}{2}\right)^{n} = 2 \qquad (1)$$

Now, we can prove easily that :

$$\sum_{n \in IN}\left(\frac{1}{2^{n^2}}\right) + \sum_{n=0}^{n=+\infty}\left(\frac{(-1)^n}{2^{n^2}}\right) = 2 \times \sum_{n \in IN}\left(\frac{1}{2^{4 \times n^2}}\right) \qquad (2)$$

We can also prove that :

$$\sum_{n\in IN}\left(\frac{1}{2^{4\times n^2}}\right)+\sum_{n=0}^{n=+\infty}\left(\frac{(-1)^n}{2^{4\times n^2}}\right)=2\times\sum_{n\in IN}\left(\frac{1}{2^{4^2\times n^2}}\right) \tag{3}$$

And also prove that :

$$\sum_{n\in IN}\left(\frac{1}{2^{4^2\times n^2}}\right)+\sum_{n=0}^{n=+\infty}\left(\frac{(-1)^n}{2^{4^2\times n^2}}\right)=2\times\sum_{n\in IN}\left(\frac{1}{2^{4^3\times n^2}}\right) \tag{4}$$

We can guess and consider this formula:

$$\sum_{n\in IN}\left(\frac{1}{2^{4^A\times n^2}}\right)+\sum_{n=0}^{n=+\infty}\left(\frac{(-1)^n}{2^{4^A\times n^2}}\right)=2\times\sum_{n\in IN}\left(\frac{1}{2^{4^A\times n^2}}\right)^4 \text{ for any natural number A.} \tag{5}$$

Now let's prove this formula by induction since we only guessed it:

This formula is correct for A=0 since we can check by an easy calculations that:

$$\sum_{n\in IN}\left(\frac{1}{2^{n^2}}\right)+\sum_{n=0}^{n=+\infty}\left(\frac{(-1)^n}{2^{n^2}}\right)=2\times\sum_{n\in IN}\left(\frac{1}{2^{4\times n^2}}\right)=2\times\sum_{n\in IN}\left(\frac{1}{2^{n^2}}\right)^4$$

We consider that this formula is correct till the natural number A. Now let's check if this formula is correct for : A+1.

We should prove for A+1 that: $$\sum_{n\in IN}\left(\frac{1}{2^{4^{A+1}\times n^2}}\right)+\sum_{n=0}^{n=+\infty}\left(\frac{(-1)^n}{2^{4^{A+1}\times n^2}}\right)=2\times\sum_{n\in IN}\left(\frac{1}{2^{4^{A+1}\times n^2}}\right)^4 \tag{6}$$

The Proof for A+1:

Let's start by separating this series $\sum_{n=0}^{n=N}\left(\frac{1}{2^{4^{A+1}\times n^2}}\right)$ into two different series with N and p are natural numbers and N>0 and N is an even number.

We have : $$\sum_{n=0}^{n=N}\left(\frac{1}{2^{4^{A+1}\times n^2}}\right)=\sum_{p=0}^{p=\frac{N}{2}}\left(\frac{1}{2^{4^{A+1}\times(2\times p)^2}}\right)+\sum_{p=0}^{p=\frac{(N-1)}{2}}\left(\frac{1}{2^{4^{A+1}\times(2\times p+1)^2}}\right) \tag{7}$$

$$\text{Hence : } \sum_{n=0}^{n=N}\left(\frac{1}{2^{4^{A+1}\times n^2}}\right)=\sum_{p=0}^{p=\frac{N}{2}}\left(\frac{1}{2^{4^{A+2}\times p^2}}\right)+\sum_{p=0}^{p=\frac{(N-1)}{2}}\left(\frac{1}{2^{4^{A+1}\times(2\times p+1)^2}}\right) \tag{8}$$

If N is a not even number, then :

$$\sum_{n=0}^{n=N}\left(\frac{1}{2^{4^{A+1}\times n^2}}\right)=\sum_{p=0}^{p=\frac{(N-1)}{2}}\left(\frac{1}{2^{4^{A+2}\times p^2}}\right)+\sum_{p=0}^{p=\frac{N}{2}}\left(\frac{1}{2^{4^{A+1}\times(2\times p+1)^2}}\right) \tag{9}$$

And let's separate this series $\sum_{n=0}^{n=N}\left(\frac{(-1)^n}{2^{4^{A+1}\times n^2}}\right)$ into two different series with N and p are natural numbers and N>0 and N is an even number .

$$\text{We have : } \sum_{n=0}^{n=N}\left(\frac{(-1)^n}{2^{4^{A+1}\times n^2}}\right)=\sum_{p=0}^{p=\frac{N}{2}}\left(\frac{(-1)^{2\times p}}{2^{4^{A+1}\times(2\times p)^2}}\right)+\sum_{p=0}^{p=\frac{(N-1)}{2}}\left(\frac{(-1)^{2\times p+1}}{2^{4^{A+1}\times(2\times p+1)^2}}\right) \tag{10}$$

$$\text{Hence : } \sum_{n=0}^{n=N}\left(\frac{(-1)^n}{2^{4^{A+1}\times n^2}}\right)=\sum_{p=0}^{p=\frac{N}{2}}\left(\frac{1}{2^{4^{A+2}\times p^2}}\right)-\sum_{p=0}^{p=\frac{(N-1)}{2}}\left(\frac{(-1)^{2\times p}}{2^{4^{A+1}\times(2\times p+1)^2}}\right) \tag{11}$$

$$\text{Consequenly : } \sum_{n=0}^{n=N}\left(\frac{(-1)^n}{2^{4^{A+1}\times n^2}}\right)=\sum_{p=0}^{p=\frac{N}{2}}\left(\frac{1}{2^{4^{A+2}\times p^2}}\right)-\sum_{p=0}^{p=\frac{(N-1)}{2}}\left(\frac{1}{2^{4^{A+1}\times(2\times p+1)^2}}\right) \tag{12}$$

If N is a not even number, then :

$$\sum_{n=0}^{n=N}\left(\frac{(-1)^n}{2^{4^{A+1}\times n^2}}\right)=\sum_{p=0}^{p=\frac{(N-1)}{2}}\left(\frac{1}{2^{4^{A+2}\times p^2}}\right)-\sum_{p=0}^{p=\frac{N}{2}}\left(\frac{1}{2^{4^{A+1}\times(2\times p+1)^2}}\right) \tag{13}$$

Now let's add the two equations :

$$\text{We get : } \sum_{n=0}^{n=N}\left(\frac{1}{2^{4^{A+1}\times n^2}}\right)+\sum_{n=0}^{n=N}\left(\frac{(-1)^n}{2^{4^{A+1}\times n^2}}\right)=2\times\sum_{p=0}^{p=\frac{N}{2}}\left(\frac{1}{2^{4^{A+2}\times p^2}}\right) \text{ for N even.} \tag{14}$$

And we get : $\sum_{n=0}^{n=N}\left(\frac{1}{2^{4^{A+1}}\times n^2}\right)+\sum_{n=0}^{n=N}\left(\frac{(-1)^n}{2^{4^{A+1}}\times n^2}\right)=2\times\sum_{p=0}^{p=\frac{N-1}{2}}\left(\frac{1}{2^{4^{A+2}}\times p^2}\right)$ for N not even. (15)

We conclude finally that : $\sum_{n=0}^{n=+\infty}\left(\frac{1}{2^{4^{A+1}}\times n^2}\right)+\sum_{n=0}^{n=+\infty}\left(\frac{(-1)^n}{2^{4^{A+1}}\times n^2}\right)=2\times\sum_{n=0}^{n=+\infty}\left(\frac{1}{2^{4^{A+2}}\times n^2}\right)$ (16)

Which means that the formula (5) is correct for A+1.

And thus the formula $\sum_{n\in IN}\left(\frac{1}{2^{4^A}\times n^2}\right)+\sum_{n=0}^{n=+\infty}\left(\frac{(-1)^n}{2^{4^A}\times n^2}\right)=2\times\sum_{n\in IN}\left(\frac{1}{2^{4^A}\times n^2}\right)^4$ is correct for any natural number A.

This is an easy proof for the recurrence formula which is very useful for the demonstrations of this article since we will use it in order to develop a new interesting and important equation of infinite series and sums.

The generalization of this recurrence formula for all natural powers C with C>1 is also easy to remark and it can be very useful :

You can verify easily for any real number B with B> 1 that :

$$\sum_{n\in IN}\left(\frac{1}{B^{8^A}\times n^3}\right)+\sum_{n=0}^{n=+\infty}\left(\frac{(-1)^n}{B^{8^A}\times n^3}\right)=2\times\sum_{n\in IN}\left(\frac{1}{B^{8^A}\times n^3}\right)^8 \quad (17)$$

And also : $\sum_{n\in IN}\left(\frac{1}{B^{16^A}\times n^4}\right)+\sum_{n=0}^{n=+\infty}\left(\frac{(-1)^n}{B^{16^A}\times n^4}\right)=2\times\sum_{n\in IN}\left(\frac{1}{B^{16^A}\times n^4}\right)^{16}$ (18)

Hence, by following the same steps of the demonstrations above concerning the recurrence relation, we can conclude with a similar recurrence proof that for any natural number A and for any natural number ***C*** with ***C>1*** and for any real number ***B*** with ***B>1*** we have:

$$\sum_{n\in IN}\left(\frac{1}{B^{2^{A\times C}}\times n^C}\right)+\sum_{n=0}^{n=+\infty}\left(\frac{(-1)^n}{B^{2^{A\times C}}\times n^C}\right)=2\times\sum_{n\in IN}\left(\frac{1}{B^{2^{A\times C}}\times n^C}\right)^{2^C} \quad (19)$$

2. The new remarked formula of series and the concluded equation :

We proved that : $\sum_{n\in IN}\left(\frac{1}{2^{4^4\times n^2}}\right)+\sum_{n=0}^{n=+\infty}\left(\frac{(-1)^n}{2^{4^4\times n^2}}\right)=2\times\sum_{n\in IN}\left(\frac{1}{2^{4^4\times n^2}}\right)^4$

Hence : $\sum_{n\in IN}\left(\frac{1}{2^{n^2}}\right)=2\times\sum_{n\in IN}\left(\frac{1}{2^{4\times n^2}}\right)-\sum_{n=0}^{n=+\infty}\left(\frac{(-1)^n}{2^{n^2}}\right)$ (20)

Now let's use this simple recurrence relation to develop a new interesting and important equation :

By applying the demonstrated formula (5) to the term : $\sum_{n\in IN}\left(\frac{1}{2^{4\times n^2}}\right)^4$ (21)

We have : $2\times\sum_{n\in IN}\left(\frac{1}{2^{4\times n^2}}\right)=2\times\left(2\times\sum_{n\in IN}\left(\frac{1}{2^{4^2\times n^2}}\right)-\sum_{n=0}^{n=+\infty}\left(\frac{(-1)^n}{2^{4\times n^2}}\right)\right)$ (22)

Consequently : $\sum_{n\in IN}\left(\frac{1}{2^{n^2}}\right)=2\times\left(2\times\sum_{n\in IN}\left(\frac{1}{2^{4^2\times n^2}}\right)-\sum_{n=0}^{n=+\infty}\left(\frac{(-1)^n}{2^{4\times n^2}}\right)\right)-\sum_{n=0}^{n=+\infty}\left(\frac{(-1)^n}{2^{n^2}}\right)$ (23)

And thus : $\sum_{n\in IN}\left(\frac{1}{2^{n^2}}\right)=\sum_{n\in IN}\left(\frac{1}{2^{4^2\times n^2-2}}\right)-\sum_{n=0}^{n=+\infty}\left(\frac{(-1)^n}{2^{n^2}}\right)-\sum_{n=0}^{n=+\infty}\left(\frac{(-1)^n}{2^{4\times n^2-1}}\right)$ (24)

We repeat this operation again, and we get :

$$\sum_{n\in IN}\left(\frac{1}{2^{n^2}}\right)=\sum_{n\in IN}\left(\frac{1}{2^{4^3\times n^2-3}}\right)-\sum_{n=0}^{n=+\infty}\left(\frac{(-1)^n}{2^{n^2}}\right)-\sum_{n=0}^{n=+\infty}\left(\frac{(-1)^n}{2^{4\times n^2-1}}\right)-\sum_{n=0}^{n=+\infty}\left(\frac{(-1)^n}{2^{4^2\times n^2-2}}\right) \quad (25)$$

We repeat this operation L times and we get with L>1:

$$\sum_{n\in IN}\left(\frac{1}{2^{n^2}}\right)=\sum_{n\in IN}\left(\frac{1}{2^{4^L\times n^2-L}}\right)-\sum_{i=0}^{i=L-1}\left(\sum_{n=0}^{n=+\infty}\left(\frac{(-1)^n}{2^{4^i\times n^2-i}}\right)\right) \quad (26)$$

Let's avoid the case of n=0 in our equation by following these steps :

we have : $\sum_{n\in IN}\left(\frac{1}{2^{n^2}}\right)=\sum_{n=1}^{n=+\infty}\left(\frac{1}{2^{n^2}}\right)+1$ (27)

And we have : $\sum_{n\in IN}\left(\frac{1}{2^{4^L\times n^2-L}}\right)=2^L+\sum_{n=1}^{n=+\infty}\left(\frac{1}{2^{4^L\times n^2-L}}\right)$ (28)

And we have : $\sum_{i=0}^{i=L-1}\left(\sum_{n=0}^{n=+\infty}\left(\frac{(-1)^n}{2^{4^i\times n^2-i}}\right)\right)=\sum_{i=0}^{i=L-1}\left(\sum_{n=1}^{n=+\infty}\left(\frac{(-1)^n}{2^{4^i\times n^2-i}}\right)\right)+\sum_{i=0}^{i=L-1}2^i$ (29)

However we know that : $2^L-1=\sum_{i=0}^{i=L-1}2^i$ (30)

Hence, the equation (26) becomes :

$$\sum_{n=1}^{n=+\infty}\left(\frac{1}{2^{n^2}}\right)=\sum_{n=1}^{n=+\infty}\left(\frac{1}{2^{4^L\times n^2-L}}\right)-\sum_{i=0}^{i=L-1}\left(\sum_{n=1}^{n=+\infty}\left(\frac{(-1)^n}{2^{4^i\times n^2-i}}\right)\right) \quad (31)$$

We know that : $\lim_{L\to+\infty} 4^L\times n^2-L=+\infty$ for any n>0 and since the sum of an infinity of zeros is a zero,

then we have : $\sum_{n\in}\lim_{L\to+\infty}\left(\frac{1}{2^{4^L\times n^2-L}}\right)=0$ (32)

Now let's investigate if this equation is correct:

$$\lim_{L\to+\infty}\sum_{n=1}^{n=+\infty}\left(\frac{1}{2^{4^L\times n^2-L}}\right)=\sum_{n=1}^{n=+\infty}\lim_{L\to+\infty}\left(\frac{1}{2^{4^L\times n^2-L}}\right)=0 \quad (33)$$

In this case, we should prove first that $\sum_{n=1}^{n=+\infty}\left(\frac{1}{2^{4^L\times n^2-L}}\right)$ is uniformly convergent .

We have : $L>\frac{\ln(L)}{\ln(2)}\Rightarrow 2\times L\times\ln(2)>2\times\ln(L)\Rightarrow\exp(2\times L\times\ln(2))>\exp(2\times\ln(L))$ (34)

Consequently : $4^L\times n^2-L>L^2\times n^2-L>L^2\times n^2-L^2$ (35)

Hence : $2^{4^L\times n^2-L}>2^{L^2\times(n^2-1)}>2^{(n^2-1)}$ (36)

And for L>1, we have :

$$2^{4^L \times n^2 - L} \geq 2^{\left(n^2-1\right)} \Rightarrow \frac{1}{2^{4^L \times n^2 - L}} \leq \frac{1}{2^{\left(n^2-1\right)}} \quad (37)$$

We conclude that : $\left|\sum_{n=k'}^{n=+\infty}\left(\frac{1}{2^{4^L \times n^2 - L}}\right)\right| \leq \sum_{n=k'}^{n=+\infty}\left|\left(\frac{1}{2^{4^L \times n^2 - L}}\right)\right| \leq \sum_{n=k'}^{n=+\infty}\left|\left(\frac{1}{2^{n^2-1}}\right)\right|$ for L >1 and for any natural number k'. (38)

However the series $\sum_{n=1}^{n=+\infty}\left(\frac{1}{2^{n^2-1}}\right)$ is absolutely convergent and $\lim_{n \to +\infty}\left(\frac{1}{2^{n^2-1}}\right) = 0$

Hence : $(\forall \varepsilon > 0), (\exists k \in IN)$ such as $\left|\frac{1}{2^{\left(k^2-1\right)}}\right| < \frac{\varepsilon}{M+k+1}$ with M>k and M natural number.

And thus : $(\forall \varepsilon > 0), (\exists k \in IN)$ $\sum_{n=k}^{n=M}\left|\frac{1}{2^{\left(k^2-1\right)}}\right| < \sum_{n=k}^{n=M}\left(\frac{\varepsilon}{M+k+1}\right) < \left(\frac{\varepsilon}{M+k+1}\right) \times (M-k+1) < \varepsilon$ (39)

since the terms of the series $\sum_{n=k}^{n=M}\left(\frac{\varepsilon}{M+k+1}\right)$ get smaller when k increases and since :

$$\left(\frac{1}{M+k+1}\right) \times (M-k+1) < 1 \quad (40)$$

We conclude that : $(\forall \varepsilon > 0), (\exists k \in IN)$ $\sum_{n=k}^{n=+\infty}\left|\frac{1}{2^{\left(k^2-1\right)}}\right| < \varepsilon$ (41)

Consequently : $(\forall \varepsilon > 0), (\exists k \in IN)$ $\left|\sum_{n=k}^{n=+\infty}\left(\frac{1}{2^{4^L \times n^2 - L}}\right)\right| < \varepsilon$ for L>1 (42)

if we consider a natural number m>k, we also have :

$$\left|\sum_{n=m}^{n=+\infty}\left(\frac{1}{2^{4^L \times n^2 - L}}\right)\right| < \varepsilon \quad \text{for L>1} \quad (43)$$

Hence, we conclude that :

$$(\forall \varepsilon > 0), (\exists k, m \in IN), with (m > k) \quad \left|\sum_{n=1}^{n=m}\left(\frac{1}{2^{4^L \times n^2 - L}}\right) - \sum_{n=1}^{n=k}\left(\frac{1}{2^{4^L \times n^2 - L}}\right)\right| < \varepsilon \quad \text{for L>1} \quad (44)$$

And thus, since k and m are both bigger than a finite natural number, than we conclude from the Cauchy Criterion in the formula (44) that:

$\sum_{n=1}^{n=+\infty}\left(\frac{1}{2^{4^L}\times n^2-L}\right)$ is uniformly convergent for L>1 since the function which is the limit of

$\sum_{n=1}^{n=+\infty}\left(\frac{1}{2^{4^L}\times n^2-L}\right)$ is obviously a positive real and finite function.

Finally, we proved that: $\lim_{L\to+\infty}\sum_{n=1}^{n=+\infty}\left(\frac{1}{2^{4^L}\times n^2-L}\right)=\sum_{n=1}^{n=+\infty}\lim_{L\to+\infty}\left(\frac{1}{2^{4^L}\times n^2-L}\right)=0$ for L>1 (45)

Consequently, when L reaches the infinity : $\sum_{n=1}^{n=+\infty}\left(\frac{1}{2^{n^2}}\right)=-\sum_{i=0}^{i=+\infty}\left(\sum_{n=1}^{n=+\infty}\left(\frac{(-1)^n}{2^{4^i}\times n^2-i}\right)\right)$ (46)

Hence : $\sum_{n=1}^{n=+\infty}\left(\frac{1}{2^{n^2}}\right)=-\sum_{i=0}^{i=+\infty}\left(2^i\times\sum_{n=1}^{n=+\infty}\left(\frac{(-1)^n}{2^{4^i}\times n^2}\right)\right)$ (47)

Let's consider F such as : $F=-\sum_{i=0}^{i=+\infty}\left(2^i\times\sum_{n=1}^{n=+\infty}\left(\frac{(-1)^n}{2^{4^i}\times n^2}\right)\right)$ and we know that F is convergent and positive.

We remark that the series $\sum_{n=1}^{n=+\infty}\left(\frac{(-1)^n}{2^{4^i}\times n^2}\right)$ is an alternating convergent series since : $\frac{1}{2^{4^i}\times n^2}$ decreases monotonically and $\lim_{n\to+\infty}\left(\frac{1}{2^{4^i}\times n^2}\right)=0$ for any natural number i.

However, the limit of $\sum_{n=1}^{n=+\infty}\left(\frac{(-1)^n}{2^{4^i}\times n^2}\right)$ depends on i.

3. The Cauchy product with F :

Now let's apply the Cauchy product by using the absolutely convergent series : $\sum_{n=0}^{n=+\infty}\left(\frac{1}{2}\right)^n = 2$

and the convergent series F since the the Cauchy product converges.

We have : $C_k = \sum_{i=0}^{i=k}\left(\frac{1}{2}\right)^{k-i} \times \left(2^i \times \sum_{n=1}^{n=+\infty}\left(\frac{(-1)^n}{2^{4^i \times n^2}}\right)\right) = -\sum_{i=0}^{i=k}\left(\left(\frac{1}{2}\right)^{k-2\times i} \times \sum_{n=1}^{n=+\infty}\left(\frac{(-1)^n}{2^{4^i \times n^2}}\right)\right)$ (48)

Hence : $C_k = -\sum_{i=0}^{i=k}\left(\left(\frac{1}{2}\right)^k \times \sum_{n=1}^{n=+\infty}\left(\frac{(-1)^n}{2^{4^i \times n^2 - 2\times i}}\right)\right) = -\left(\frac{1}{2}\right)^k \times \sum_{i=0}^{i=k}\left(\sum_{n=1}^{n=+\infty}\left(\frac{(-1)^n}{2^{4^i \times n^2 - 2\times i}}\right)\right)$ (49)

Consequently : $F \times \sum_{n\in IN}\left(\frac{1}{2}\right)^n = 2\times F = -\lim_{k\to+\infty}\left(\left(\frac{1}{2}\right)^k \times \sum_{i=0}^{i=k}\left(\sum_{n=1}^{n=+\infty}\left(\frac{(-1)^n}{2^{4^i \times n^2 - 2\times i}}\right)\right)\right)$ (50)

And we know that : $-\lim_{k\to+\infty}\left(\frac{1}{2}\right)^k = 0$ Hence : $\sum_{i=0}^{i=+\infty}\left(\sum_{n=1}^{n=+\infty}\left(\frac{(-1)^n}{2^{4^i \times n^2 - 2\times i}}\right)\right) = -\infty$ (51)

And also, we remark that the series $\sum_{n=1}^{n=+\infty}\left(\frac{(-1)^n}{2^{4^i \times n^2 - 2\times i}}\right)$ is an alternating convergent series since :

$\frac{1}{2^{4^i \times n^2 - 2\times i}}$ decreases monotonically and $\lim_{n\to+\infty}\left(\frac{1}{2^{4^i \times n^2 - 2\times i}}\right) = 0$ for any natural number i.

However, the limit of $\sum_{n=1}^{n=+\infty}\left(\frac{(-1)^n}{2^{4^i \times n^2 - 2\times i}}\right)$ depends on i, and since : $\sum_{i=0}^{i=+\infty}\left(\sum_{n=1}^{n=+\infty}\left(\frac{(-1)^n}{2^{4^i \times n^2 - 2\times i}}\right)\right) = -\infty$,

then, depending on the changing i, each sum of the series $\sum_{n=1}^{n=+\infty}\left(\frac{(-1)^n}{2^{4^i \times n^2 - 2\times i}}\right)$ is either negative or positive.

We proved that : $F = -\lim_{k\to+\infty}\left(\left(\frac{1}{2}\right)^{k+1} \times \sum_{i=0}^{i=k}\left(\sum_{n=1}^{n=+\infty}\left(\frac{(-1)^n}{2^{4^i \times n^2 - 2\times i}}\right)\right)\right)$ (52)

Finally, we conclude this equation :

$$\sum_{n=1}^{n=+\infty}\left(\frac{1}{2^{n^2}}\right)=-\sum_{i=0}^{i=+\infty}\sum_{n=1}^{n=+\infty}\left(\frac{(-1)^n}{2^{4^i\times n^2-i}}\right)=-\lim_{k\to+\infty}\left(\left(\frac{1}{2}\right)^{k+1}\times\sum_{i=0}^{i=k}\left(\sum_{n=1}^{n=+\infty}\left(\frac{(-1)^n}{2^{4^i\times n^2-2\times i}}\right)\right)\right) \quad (53)$$

* The notation «lim » is maintained in the formula (53) because the increasing variable k should tend to the same value for all the terms even in the « infinity».

Many conclusions can be made from this equation, and we can also remark this equation :

$$\lim_{k\to+\infty}\sum_{i=0}^{i=k}\sum_{n=1}^{n=+\infty}\left(\frac{(-1)^n}{2^{4^i\times n^2-2\times i}}\right)=\lim_{k\to+\infty}(2)^{k+1}\times\sum_{i=0}^{i=+\infty}\sum_{n=1}^{n=+\infty}\left(\frac{(-1)^n}{2^{4^i\times n^2-i}}\right) \quad (54)$$

* The notation «lim » is maintained in the formula (54) because the increasing variable k should tend to the same value for all the terms even in the « infinity».

4. The concluded equation of infinite series and sums :

Let's also remember that by using this formula of Poisson :

$$\sum_{n\in Z}\exp\left(-\pi\times n^2\times s\right)=\frac{1}{\sqrt{s}}\times\sum_{n\in Z}\exp\left(\frac{-\pi\times n^2}{s}\right) \qquad (\text{ for any : } s>0) \quad (55)$$

We could prove in a previous short article [1] that for any real number u with $0<u<1$, we have :

$$\sum_{n\in IN}u^{\pi\times n^2}=\frac{1}{2\times\sqrt{-\ln(u)}}\times\left(\sum_{n\in Z}\exp\left(\frac{\pi\times n^2}{\ln(u)}\right)\right)+\frac{1}{2} \quad (56)$$

And by considering that : $u=\left(\frac{1}{2}\right)^{\frac{1}{\pi}}$ we get this important formula :

$$\sum_{n\in IN}\left(\frac{1}{2}\right)^{n^2}=\frac{1}{2\times\sqrt{\frac{\ln(2)}{\pi}}}\times\left(\sum_{n\in Z}\exp\left(\frac{-\pi^2\times n^2}{\ln(2)}\right)\right)+\frac{1}{2} \quad (57)$$

Hence : $\sum_{n=1}^{n=+\infty}\left(\frac{1}{2}\right)^{n^2} = \frac{1}{2\times\sqrt{\frac{\ln(2)}{\pi}}}\times\left(\sum_{n\in Z}\exp\left(\frac{-\pi^2\times n^2}{\ln(2)}\right)\right)-\frac{1}{2}$ (58)

Finally, we could prove this fundamental equation of four different infinite series and sums :

$$\sum_{n=1}^{n=+\infty}\left(\frac{1}{2^{n^2}}\right) = -\sum_{i=0}^{i=+\infty}\sum_{n=1}^{n=+\infty}\left(\frac{(-1)^n}{2^{4^i\times n^2-i}}\right) = \frac{1}{2\times\sqrt{\frac{\ln(2)}{\pi}}}\times\left(\sum_{n\in Z}\exp\left(\frac{-\pi^2\times n^2}{\ln(2)}\right)\right)-\frac{1}{2} \quad (59)$$

Without forgetting this important equation:

$$\sum_{n=1}^{n=+\infty}\left(\frac{1}{2^{n^2}}\right) = -\lim_{k\to+\infty}\left(\left(\frac{1}{2}\right)^{k+1}\times\sum_{i=0}^{i=k}\left(\sum_{n=1}^{n=+\infty}\left(\frac{(-1)^n}{2^{4^i\times n^2-2\times i}}\right)\right)\right) \quad (60)$$

* The notation «lim » is maintained in the formula (60) because the increasing variable k should tend to the same value for all the terms even in the « infinity».

Many important conclusions and mathematical developments can be done further by generalizing this article equation developed by the example of the studied element $q=\frac{1}{2}$ to all real numbers $q=\frac{1}{B}$ with $B\geq 2$ in order to use the element $\frac{1}{B}$ in the equations instead of $\frac{1}{2}$. This will allow us to consider that the equations of sums are equations of functions of the real numbers B.

5. Generalizing the equation of infinite series and sums to all real numbers B with $B\geq 2$ and to all natural powers C with C>1:

We can prove that for any natural number ***A*** and for any natural number ***C*** with ***C>1*** and for any real number B with B>1 we have this recurrence relation:

$$\sum_{n\in IN}\left(\frac{1}{B^{2^{A\times C}\times n^C}}\right)+\sum_{n=0}^{n=+\infty}\left(\frac{(-1)^n}{B^{2^{A\times C}\times n^C}}\right) = 2\times\sum_{n\in IN}\left(\frac{1}{B^{2^{A\times C}\times n^C}}\right)^{2^C}$$

By using this general recurrence relation L times, like in the formula (26), we can get :

$$\sum_{n\in IN}\left(\frac{1}{B^{n^C}}\right)=\sum_{n\in IN}\left(\frac{1}{\left(B^{2^{C\times L}\times n^C}\times 2^{-L}\right)}\right)-\sum_{i=0}^{i=L-1}\left(\sum_{n=0}^{n=+\infty}\left(\frac{(-1)^n}{\left(B^{2^{C\times i}\times n^C}\times 2^{-i}\right)}\right)\right) \quad \text{with } L>1 \tag{61}$$

We can prove for any real number B with $B\geq 2$ that :

$$B^{2^{C\times L}\times n^C}\times 2^{-L}\geq 2^{2^{C\times L}\times n^C}\times 2^{-L}\geq 2^{2^{C\times L}\times n^C-L} \tag{62}$$

And we know that : $\lim_{L\to+\infty} 2^{C\times L}\times n^C-L=+\infty$ for n>0 and since the sum of an infinity of zeros is a zero, then we have : $\sum_{n=1}^{n=+\infty}\lim_{L\to+\infty}\left(\frac{1}{\left(B^{2^{C\times L}\times n^C}\times 2^{-L}\right)}\right)=0$ with L>1 and $B\geq 2$ (63)

Furthermore, by following the same steps of the demonstrations above, we can prove easily that : $\sum_{n=1}^{n=+\infty}\left(\frac{1}{\left(B^{2^{C\times L}\times n^C}\times 2^{-L}\right)}\right)$ ***is uniformly convergent for any real number B with*** $B\geq 2$ ***and L>1 and for any natural number C with C>1 because :***

we have : $B^{2^{C\times L}\times n^C}\times 2^{-L}\geq 2^{2^{C\times L}\times n^C-L}\geq 2^{4^L\times n^2-L}$ (64)

Hence, from the formula (36), we conclude that : $B^{2^{C\times L}\times n^C}\times 2^{-L}>2^{L^2\times\left(n^2-1\right)}>2^{\left(n^2-1\right)}$ (65)

And also: $\left|\sum_{n=k'}^{n=+\infty}\left(\frac{1}{\left(B^{2^{C\times L}\times n^C}\times 2^{-L}\right)}\right)\right|\leq\sum_{n=k'}^{n=+\infty}\left|\left(\frac{1}{\left(B^{2^{C\times L}\times n^C}\times 2^{-L}\right)}\right)\right|\leq\sum_{n=k'}^{n=+\infty}\left|\left(\frac{1}{2^{n^2-1}}\right)\right|$ for L >1 and for any natural number k'. (66)

Consequently, from the formula (41), we conclude that :

$$(\forall\varepsilon>0),(\exists k\in IN)\ \left|\sum_{n=k}^{n=+\infty}\left(\frac{1}{\left(B^{2^{C\times L}\times n^C}\times 2^{-L}\right)}\right)\right|<\varepsilon \quad \text{for } L>1 \tag{67}$$

And thus, thanks to the Cauchy Criterion, we conclude that : $\sum_{n=1}^{n=+\infty}\left(\frac{1}{\left(B^{2^{C\times L}\times n^C}\times 2^{-L}\right)}\right)$ is uniformly convergent for ***L>1*** since the function which is the limit of $\sum_{n=1}^{n=+\infty}\left(\frac{1}{\left(B^{2^{C\times L}\times n^C}\times 2^{-L}\right)}\right)$ is obviously a positive real and finite function.

Finally, we proved that: $\lim_{L\to+\infty}\sum_{n=1}^{n=+\infty}\left(\frac{1}{\left(B^{2^{C\times L}\times n^C}\times 2^{-L}\right)}\right)=\sum_{n=1}^{n=+\infty}\lim_{L\to+\infty}\left(\frac{1}{\left(B^{2^{C\times L}\times n^C}\times 2^{-L}\right)}\right)=0$ for L>1 (68)

Consequently, for any real number ***B*** with $B\geq 2$ and for any natural number ***C*** with ***C>1***, when the number ***L*** reaches the infinity, we have :

$$\sum_{n=1}^{n=+\infty}\left(\frac{1}{B^{n^C}}\right)=-\sum_{i=0}^{i=+\infty}\left(\sum_{n=1}^{n=+\infty}\left(\frac{(-1)^n}{\left(B^{2^{C\times i}\times n^C}\times 2^{-i}\right)}\right)\right) \quad (69)$$

And by using the Cauchy product like in the demonstrations above, we conclude that :

$$\sum_{n=1}^{n=+\infty}\left(\frac{1}{B^{n^C}}\right)=\frac{1-B}{B}\times\lim_{k\to+\infty}\left(\left(\frac{1}{B}\right)^k\times\sum_{i=0}^{i=k}\left(\sum_{n=1}^{n=+\infty}\left(\frac{(-1)^n}{\left(B^{2^{C\times i}\times n^C-i}\times 2^{-i}\right)}\right)\right)\right) \quad (70)$$

* The notation «lim » is maintained in the formula (70) because the increasing variable k should tend to the same value for all the terms even in the « infinity».

Finally, we conclude the following equation for any real number B with $B\geq 2$ ***and for any natural number C with C>1 :***

$$\sum_{n=1}^{n=+\infty}\left(\frac{1}{B^{n^C}}\right)=-\sum_{i=0}^{i=+\infty}\left(\sum_{n=1}^{n=+\infty}\left(\frac{(-1)^n}{\left(B^{2^{C\times i}\times n^C}\times 2^{-i}\right)}\right)\right)=\frac{1-B}{B}\times\lim_{k\to+\infty}\left(\left(\frac{1}{B}\right)^k\times\sum_{i=0}^{i=k}\left(\sum_{n=1}^{n=+\infty}\left(\frac{(-1)^n}{\left(B^{2^{C\times i}\times n^C-i}\times 2^{-i}\right)}\right)\right)\right)$$

(71)

* The notation «lim » is maintained in the formula (71) because the increasing variable k should tend to the same value for all the terms even in the « infinity».

*Without forgetting the remark that for any real number $B\geq 2$ and for any natural number C with C>1, we have always: $\sum_{i=0}^{i=+\infty}\left(\sum_{n=1}^{n=+\infty}\left(\frac{(-1)^n}{\left(B^{2^{C\times i}\times n^C-i}\times 2^{-i}\right)}\right)\right)=-\infty$ (72)

The equation (71) has been tested succesfully on the softwares : « GEOGEBRA » and « MATLAB ».

6. An additional infinite series for the equation of infinite series and sums for all real numbers B with $B \geq 2$ and for all natural powers C with C>1:

We know that there is a power series representation for : $\dfrac{x}{1+x^2}$ that will be very useful in this article.

The power series representation is : $\dfrac{x}{1+x^2} = \sum_{i=0}^{+\infty} (-1)^i x^{2i+1}$ where : ***-1<x<1*** (73)

And we have : $(x-1)^2 \geq 0 \Rightarrow x^2 - 2x + 1 \geq 0 \Rightarrow \dfrac{x}{1+x^2} \leq \dfrac{1}{2}$ (74)

Let's consider that ***x>0*** and let's consider that : $\dfrac{x}{1+x^2} = \dfrac{1}{B^{n^c}}$ where ***B*** is a real number and where ***n*** and ***c*** are natural numbers with ***n>0*** and ***c>1***.

We remark that we have always : $B^{n^c} \geq 2$. (75)

And thus, we have : $B \geq 2$. (76)

We have also : $\dfrac{x}{1+x^2} = \dfrac{1}{B^{n^c}} \Rightarrow x^2 - B^{n^c} \times x + 1 = 0$ (77)

This equation has two solutions which are : $x_1 = \dfrac{B^{n^c} + \sqrt{B^{2\times n^c} - 4}}{2}$ and $x_2 = \dfrac{B^{n^c} - \sqrt{B^{2\times n^c} - 4}}{2}$ (78)

However, since we have ***x<1***, and since $x_1 = \dfrac{B^{n^c} + \sqrt{B^{2\times n^c} - 4}}{2}$ then we keep only one solution which is : $x = x_2 = \dfrac{B^{n^c} - \sqrt{B^{2\times n^c} - 4}}{2}$

And thus, we get this series for any real number ***B*** with $B \geq 2$ and for any natural numbers ***n*** and ***c*** with ***n>0*** and ***c>1 :***

$$\frac{1}{B^{n^c}} = \sum_{i=0}^{+\infty} (-1)^i \left(\frac{B^{n^c} - \sqrt{B^{2\times n^c} - 4}}{2} \right)^{2i+1} = \sum_{i=1}^{+\infty} (-1)^{i-1} \left(\frac{B^{n^c} - \sqrt{B^{2\times n^c} - 4}}{2} \right)^{2i-1} \quad (79)$$

Consequently, we have the following series for any real number ***B*** with $B \geq 2$ and for any natural numbers ***n*** and ***c*** with ***n>0*** and ***c>1 :***

$$\sum_{n=1}^{+\infty}\frac{1}{B^{n^c}}=\sum_{n=1}^{+\infty}\sum_{i=1}^{+\infty}(-1)^{i-1}\left(\frac{B^{n^c}-\sqrt{B^{2\times n^c}-4}}{2}\right)^{2i-1} \tag{80}$$

We proved above the following series for any real number ***B*** with $B \geq 2$ and for any natural number ***c*** with ***c>1 :***

$$\sum_{j=1}^{j=+\infty}\left(\frac{1}{B^{j^C}}\right)=-\sum_{n=0}^{n=+\infty}\left(\sum_{i=1}^{i=+\infty}\left(\frac{(-1)^i}{\left(B^{2^{C\times n}\times i^C}\times 2^{-n}\right)}\right)\right) \tag{81}$$

This means that : $$\sum_{j=1}^{j=+\infty}\left(\frac{1}{B^{j^C}}\right)=\sum_{n=1}^{n=+\infty}\sum_{i=1}^{i=+\infty}\left(\frac{(-1)^{i+1}}{\left(B^{2^{C\times(n-1)}\times i^C}\times 2^{1-n}\right)}\right) \tag{82}$$

And thus, we have the following equation for any real number ***B*** with $B \geq 2$ and for any natural number ***c*** with ***c>1 :***

$$\sum_{n=1}^{+\infty}\sum_{i=1}^{+\infty}(-1)^{i-1}\left(\frac{B^{n^c}-\sqrt{B^{2\times n^c}-4}}{2}\right)^{2i-1}=\sum_{n=1}^{n=+\infty}\sum_{i=1}^{i=+\infty}\left(\frac{(-1)^{i+1}}{\left(B^{2^{C\times(n-1)}\times i^C}\times 2^{1-n}\right)}\right) \tag{83}$$

We remark that $(-1)^{i-1}\left(\frac{B^{n^c}-\sqrt{B^{2\times n^c}-4}}{2}\right)^{2i-1}$ and $\frac{(-1)^{i+1}}{\left(B^{2^{C\times(n-1)}\times i^C}\times 2^{1-n}\right)}$ are two similar alternating sequences. Hence, let's check if the two sequences are equal or equivalent.

If our two alternating sequences are equal or equivalent, then this means that we have the following equation for any real number ***B*** with $B \geq 2$ and for any natural number ***c*** with ***c>1*** and for any natural numbers ***i*** and ***n*** with ***i>0*** and ***n>0 :***

$$(-1)^{i-1}\left(\frac{B^{n^c}-\sqrt{B^{2\times n^c}-4}}{2}\right)^{2i-1}=\frac{(-1)^{i+1}}{\left(B^{2^{C\times(n-1)}\times i^C}\times 2^{1-n}\right)} \tag{84}$$

Consequently, we get this equation for any real number ***B*** with $B \geq 2$ and for any natural number ***c*** with ***c>1*** and for any natural numbers ***i*** and ***n*** with ***i>0*** and ***n>0 :***

$$\left(\frac{B^{n^C} - \sqrt{B^{2\times n^C} - 4}}{2}\right)^{2i-1} = \frac{1}{\left(B^{2^{C\times(n-1)}\times i^C} \times 2^{1-n}\right)} \tag{85}$$

However, for ***n=1*** and ***i=2***, we get the following equation : $\left(\frac{B - \sqrt{B^2 - 4}}{2}\right)^3 = \frac{1}{B^{2^C}}$ (86)

And this is a contradictory impossible result since $\frac{1}{B^{2^C}}$ can change by changing the number ***c***,

whereas $\left(\frac{B - \sqrt{B^2 - 4}}{2}\right)^3$ never changes by changing the number ***c***.

And thus, the two infinite series with alternating general terms which

are : $\sum_{n=1}^{+\infty}\sum_{i=1}^{+\infty}(-1)^{i-1}\left(\frac{B^{n^C} - \sqrt{B^{2\times n^C} - 4}}{2}\right)^{2i-1}$ and $\sum_{n=1}^{n=+\infty}\sum_{i=1}^{i=+\infty}\left(\frac{(-1)^{i+1}}{\left(B^{2^{C\times(n-1)}\times i^C} \times 2^{1-n}\right)}\right)$ are indeed two different series that have the same limit at the infinity.

We conclude that the infinite series $\sum_{n=1}^{+\infty}\sum_{i=1}^{+\infty}(-1)^{i-1}\left(\frac{B^{n^C} - \sqrt{B^{2\times n^C} - 4}}{2}\right)^{2i-1}$ is also a new infinite series equivalent to $\sum_{n=1}^{+\infty}\frac{1}{B^{n^C}}$.

7. Conclusions regarding Theta Functions in their « Nullwert » forms:

We remind that we can link directly the Theta-3 Functions to this article by considering the transendence findings of the Dr Daniel Bertrand in his work 'Theta functions and transcendence' [4] and by respecting that:

$$\theta_3(q) = 1 + 2\times\sum_{n\geq 1} q^{n^2} = 1 + 2\times\sum_{n\geq 1}\frac{1}{B^{n^2}} \quad \text{with : } B = \frac{1}{q} \tag{87}$$

A short article proves that Theta-3 Functions give irrational outputs if $B \geq 2$ when the inputs are rational [2] and an article of professor Bertrand [4] proves that $\sum_{n \in IN} \left(\frac{1}{B}\right)^{n^2}$ is not rational unless ***B*** is a transcendental number.

And from this article [4], we can also use the following representation for Theta-4 Functions :

$$\theta_4(q) = 1 + 2 \times \sum_{n \geq 1} (-1)^n \times q^{n^2} = 1 + 2 \times \sum_{n \geq 1} \frac{(-1)^n}{B^{n^2}} \qquad \text{with : } B = \frac{1}{q} \tag{88}$$

And from this equation proved in our article: $\sum_{n=1}^{n=+\infty} \left(\frac{1}{B^{n^C}}\right) = -\sum_{i=0}^{i=+\infty} \sum_{n=1}^{n=+\infty} \left(\frac{(-1)^n}{\left(B^{2^{C \times i} \times n^C} \times 2^{-i}\right)}\right)$, we remark that we have :

$$\sum_{n=1}^{n=+\infty} \left(\frac{1}{B^{n^2}}\right) + \sum_{n=1}^{n=+\infty} \left(\frac{(-1)^n}{B^{n^2}}\right) = -\sum_{i=1}^{i=+\infty} \sum_{n=1}^{n=+\infty} \left(\frac{(-1)^n}{\left(B^{2^{2 \times i} \times n^2} \times 2^{-i}\right)}\right) \tag{89}$$

And thus, we have : $2 + 2 \times \sum_{n=1}^{n=+\infty} \left(\frac{1}{B^{n^2}}\right) + 2 \times \sum_{n=1}^{n=+\infty} \left(\frac{(-1)^n}{B^{n^2}}\right) = 2 - \sum_{i=1}^{i=+\infty} \sum_{n=1}^{n=+\infty} \left(\frac{(-1)^n \times 2^{i+1}}{\left(B^{4^i \times n^2}\right)}\right)$ (90)

We could consequently prove this interesting equation for Theta Functions:

$$\theta_3(q) + \theta_4(q) = 2 - \sum_{i=1}^{i=+\infty} \sum_{n=1}^{n=+\infty} \left((-1)^n \times 2^{i+1} \times q^{4^i \times n^2}\right) \text{ where } q \leq \frac{1}{2} \tag{91}$$

Furthermore, according to the Jacobi Identity, we can use the following transformation:

$$\theta_3(q^4) = \frac{1}{2} \times \theta_3(q) + \frac{1}{2} \times \theta_4(q) = 1 - \sum_{i=1}^{i=+\infty} \sum_{n=1}^{n=+\infty} \left((-1)^n \times 2^i \times q^{4^i \times n^2}\right) \text{ where } q \leq \frac{1}{2} \tag{92}$$

And by substituting the number $\boldsymbol{q}$ by the number $\boldsymbol{q'}$ where we have : $q'=q^4\leq\frac{1}{16}$, we remark that :

$$\theta_3(q')=1-\sum_{i=1}^{i=+\infty}\sum_{n=1}^{n=+\infty}\left((-1)^n\times 2^i\times q'^{4^{i-1}\times n^2}\right)\text{ where }q'\leq\frac{1}{16} \quad (93)$$

Finally we conclude our last equation of four infinite series and sums for Theta-3 Functions and any real number $\boldsymbol{q}$:

$$\theta_3(q)=1+2\times\sum_{n\geq 1}q^{n^2}=1-\sum_{i=0}^{i=+\infty}\left(\sum_{n=1}^{n=+\infty}\left((-1)^n\times q^{4^i\times n^2}\times 2^{i+1}\right)\right)=1+2\times\sum_{n=1}^{+\infty}\sum_{i=1}^{+\infty}(-1)^{i-1}\left(\frac{q^{-n^2}-\sqrt{q^{-2\times n^2}-4}}{2}\right)^{2i-1}$$

$$\text{where }q\leq\frac{1}{2} \quad (94)$$

And : $\theta_3(q)=1-\sum_{i=1}^{i=+\infty}\sum_{n=1}^{n=+\infty}\left((-1)^n\times 2^i\times q^{4^{i-1}\times n^2}\right)$ where $q\leq\frac{1}{16}$ (95)

8. Summary and conclusion :

We could guess and use a generalized recurrence relation which is

$\sum_{n\in IN}\left(\frac{1}{B^{2^{A\times C}\times n^C}}\right)+\sum_{n=0}^{n=+\infty}\left(\frac{(-1)^n}{B^{2^{A\times C}\times n^C}}\right)=2\times\sum_{n\in IN}\left(\frac{1}{B^{2^{A\times C}\times n^C}}\right)^{2^C}$ in order to prove an equation of infinite series and sum that are equivalent to the case of $\sum_{n=1}^{+\infty}\frac{1}{2^{n^2}}$.

We also generalized this result in order to prove an equation of infinite series and sums that are equivalent to $\sum_{n=1}^{+\infty}\frac{1}{B^{n^c}}$ for any real number $\boldsymbol{B}$ with $B\geq 2$.

We ended then the article by using a power series representation for : $\frac{x}{1+x^2}$ in order to add a new infinite series as an element of our equation of three infinite series and sums.

Afterwards, we could conclude this equation for any real number $\boldsymbol{B}$ with $B \geq 2$ and for any natural number $\boldsymbol{C}$ with $\boldsymbol{C>1}$:

$$\sum_{n=1}^{n=+\infty}\left(\frac{1}{B^{n^C}}\right)=-\sum_{i=0}^{i=+\infty}\left(\sum_{n=1}^{n=+\infty}\left(\frac{(-1)^n}{\left(B^{2^{C\times i}\times n^C}\times 2^{-i}\right)}\right)\right)=\sum_{n=1}^{+\infty}\sum_{i=1}^{+\infty}(-1)^{i-1}\left(\frac{B^{n^C}-\sqrt{B^{2\times n^C}-4}}{2}\right)^{2i-1}$$

Without forgetting this result : $\sum_{n=1}^{n=+\infty}\left(\frac{1}{B^{n^C}}\right)=\frac{1-B}{B}\times\lim_{k\to+\infty}\left(\left(\frac{1}{B}\right)^k\times\sum_{i=0}^{i=k}\left(\sum_{n=1}^{n=+\infty}\left(\frac{(-1)^n}{\left(B^{2^{C\times i}\times n^C-i}\times 2^{-i}\right)}\right)\right)\right)$

Where the notation «lim » is maintained because the increasing variable $\boldsymbol{k}$ should tend to the same value for all the terms even in the « infinity».

And let's remember that for any real number $B \geq 2$ and for any natural number $\boldsymbol{C}$ with $\boldsymbol{C>1}$, we have always: $\sum_{i=0}^{i=+\infty}\left(\sum_{n=1}^{n=+\infty}\left(\frac{(-1)^n}{\left(B^{2^{C\times i}\times n^C-i}\times 2^{-i}\right)}\right)\right)=-\infty$.

Finally, after that these results have been tested successfully on the softwares : *« **GEOGEBRA** »* and *« **MATLAB** »,* new useful and simple transformations of Theta-3 and Theta-4 functions in their « Nullwert » forms have been also proven. We could consequently prove this interesting equation for Theta-3 Functions:

$$\theta_3(q)+\theta_4(q)=2-\sum_{i=1}^{i=+\infty}\sum_{n=1}^{n=+\infty}\left((-1)^n\times 2^{i+1}\times q^{4^i\times n^2}\right)\text{ where } q\leq\frac{1}{2}$$

Furthermore, thanks to the Jacobi Identity, we could use the following transformation :

$$\theta_3\left(q^4\right)=\frac{1}{2}\times\theta_3(q)+\frac{1}{2}\times\theta_4(q)=1-\sum_{i=1}^{i=+\infty}\sum_{n=1}^{n=+\infty}\left((-1)^n\times 2^i\times q^{4^i\times n^2}\right)\text{ where } q\leq\frac{1}{2}$$

And by substituting the number $\boldsymbol{q}$ by the number $\boldsymbol{q'}$ where we have : $q'=q^4\leq\frac{1}{16}$, we could remark that : $\theta_3(q')=1-\sum_{i=1}^{i=+\infty}\sum_{n=1}^{n=+\infty}\left((-1)^n\times 2^i\times q'^{4^{i-1}\times n^2}\right)$ where $q'\leq\frac{1}{16}$

Finally we concluded our last equation of four infinite series and sums for Theta-3 Functions and any real number ***q***:

$$\theta_3(q)=1+2\times\sum_{n\geq1}q^{n^2}=1-\sum_{i=0}^{i=+\infty}\left(\sum_{n=1}^{n=+\infty}\left((-1)^n\times q^{4^i\times n^2}\times 2^{i+1}\right)\right)=1+2\times\sum_{n=1}^{+\infty}\sum_{i=1}^{+\infty}(-1)^{i-1}\left(\frac{q^{-n^2}-\sqrt{q^{-2\times n^2}-4}}{2}\right)^{2i-1}$$

where $q\leq\frac{1}{2}$

And : $\theta_3(q)=1-\sum_{i=1}^{i=+\infty}\sum_{n=1}^{n=+\infty}\left((-1)^n\times 2^i\times q^{4^{i-1}\times n^2}\right)$ where $q\leq\frac{1}{16}$

We remak also that since we have: $\theta_3(q)+\sum_{i=1}^{i=+\infty}\sum_{n=1}^{n=+\infty}\left((-1)^n\times 2^i\times q^{4^{i-1}\times n^2}\right)=1$ then the series of the forms

$\theta_3(q)+\sum_{i=1}^{i=+\infty}\sum_{n=1}^{n=+\infty}\left((-1)^n\times 2^i\times q^{4^{i-1}\times n^2}\right)$ can be very useful in probability and statistics ditributions.

And we remind again from the previous article [1] this useful result by the formula of Poisson :

$$\theta_3(q)=\frac{1}{\sqrt{\frac{-\ln(q)}{\pi}}}\times\left(\sum_{n\in Z}\exp\left(\frac{\pi^2\times n^2}{\ln(q)}\right)\right) \qquad \text{where } q\leq\frac{1}{2}$$

At the end, the readers are invited to develop and conclude other results related to the complex versions of Theta Functions and Zeta functions by using the formulas of series and transformations of the « Nullwert » forms of Theta Functions in this work. And thus, we hope that this article will be useful for simplifying many problems related to series, to number theory and even to calculus in general.

Declarations:

Ethical Approval

this declaration is “not applicable”.

Competing interests

this declaration is “not applicable”.

Authors' contributions

this declaration is “not applicable”.

Funding

this declaration is “not applicable”.

Availability of data and materials

this declaration is “not applicable”.

References :

[1] Akram Louiz. A discussion of rationality and continuity related to Theta Functions by using the summation of Poisson. Asia Mathematika, In press, ⟨10.5281/zenodo.10609879⟩. ⟨hal-04442443⟩

[2] Louiz , A. (2024). Irrational Outputs from Rational Inputs in Theta-3 Functions Using Dirichlet’s Theorem . General Letters in Mathematics, 14 (3), 56-62, 10.31559/glm2024.14.3.2

[3] ALVES, F. R. (2014). Visualizing the behavior of infinite series and complex power series with the GeoGebra. *GeoGebra International Journal of Romania*, *4*(1), 1-10.

[4] Bertrand, D. (1997). Theta functions and transcendence. The Ramanujan Journal, 1(4), 339-350.

[5] Igusa, J. I. (2012). *Theta functions* (Vol. 194). Springer Science & Business Media.

[6] Dubrovin, B. A. (1981). Theta functions and non-linear equations. *Russian mathematical surveys*, *36*(2), 11.

[7] Wei, J., Geng, X., & Zeng, X. (2019). The Riemann theta function solutions for the hierarchy of Bogoyavlensky lattices. *Transactions of the American Mathematical Society*, *371*(2), 1483-1507.

FSC
www.fsc.org
MIX
Papier | Fördert
gute Waldnutzung
FSC® C083411

Kolibri 360 GmbH
Ferdinand-Jühlke-Straße 7
99095 Erfurt, Deutschland
produktsicherheit@kolibri360.de

Druck:
CPI Druckdienstleistungen GmbH
im Auftrag der
Kolibri 360 GmbH
Ferdinand-Jühlke-Str. 7
99095 Erfurt